Praise for Alan Boss's *THE CROWDED UNIVERSE*

"In this short and lucid review of his field, [Boss] traces the developments of the last fifteen years in chronological, diarylike entries, so that we can share with him the excitement of discovery. . . . The tone of Boss's book, accordingly, is excited and hopeful, but there's also a note of wry irony in his descriptions of the political trials astronomers have gone through trying to promote their research. And despite the successes of the past decade, Boss senses that it may be increasingly difficult for astronomers to attract the sums needed to continue the search for habitable planets. Readers of this book, I am certain, will hope his fears are unsubstantiated." —*Natural History*

"Alan Boss . . . weaves the story of Kepler with the larger tale of the booming field of exoplanets. As someone whose career in astronomy has spanned the period Boss discusses, I am glad someone was taking notes. It is fun to revisit the days when each new planetary discovery was an exciting event. Multiple teams struggled to outdo the others with firsts. First planet at the distance of the Earth! First transiting planet! First multiple planet system! It is easy to forget that most of the exoplanet field is less than a decade old." —Michael Brown, *Nature*

"If Alan Boss's excellent new book is anything to go by, the next few years could see some dramatic revelations about our cosmic neighbourhood. . . . In *The Crowded Universe* he skillfully recounts how astronomers have gradually become better acquainted with the exoplanets—planets orbiting stars other than the Sun." —*BBC Focus Magazine*

"[Boss] allows readers to experience the nature of scientific life, with all the setbacks, blind alleys and complex human relationships that make the process of discovery anything but linear."

—*Milwaukee Journal Sentinel*

"Readers will appreciate Boss' approach if they love science and technology for open questions, possibilities and problems rather than for answers and solutions." —*Arkansas Democrat Gazette*

"[*The Crowded Universe*] is a stunning story, recasting scientists as detectives developing and using new tools to expand knowledge of our exciting universe. Scientist Alan Boss . . . has found a second career as an interpreter of the scientific enterprise for the general public."

—*Space Times*

"What makes the book useful is that it ties together not just the science but also the technical and political issues that make it possible (or, sometimes, impossible) to perform science. . . . Offer[s] a good way to quickly get up to speed on the state of and prospects for exoplanet research."

—*The Space Review*

"[A] fresh look at the on-going search to scope out Earth-like planets. . . . [Boss's] writing style is humorous as well as enlightening."

—*Space Coalition Blog*

"Solid coverage of one of the most exciting topics in science."

—*Kirkus Reviews*

"[T]he book reads like an adventure yarn, reminiscent of archaeologists looking for fabled lost cities. . . . [A] thoroughly fascinating account."

—*Choice*

"The search for life beyond the Earth, and the study of planets orbiting other stars, are surely among the most fascinating topics in 21st century science. Alan Boss offers a clear and masterly guide to these exciting and fast-moving subjects."

—Sir Martin Rees, Cambridge Institute of Astronomy

"The search for planets outside our solar system has become a cottage industry. In *The Crowded Universe*, Alan Boss weaves a 'you are there' narrative that reaches behind the scenes of this thrilling new field, exposing the reader to the people, the politics, and the sheer joy of doing science."

—Neil deGrasse Tyson, Astrophysicist, American Museum of Natural History, and author of *The Pluto Files*

"*The Crowded Universe* is a thorough depiction of the events and people involved in one of the greatest milestones in the history of science: the detection of other planetary systems in the Milky Way. The author is one of the primary players in this ongoing saga, and he tells the story with commendable detail. If you want to see how science works at its best, read this book."

—Dr. Frank Drake, Director, Carl Sagan Center for the Study of Life in the Universe, SETI Institute

"Rarely is the history of science so accurately told as in this lively and authoritative book. Alan Boss offers insights about our terrestrial origins, our extraterrestrial brethren, and our destiny in the Galaxy, placing our Earth in the cosmic context for the first time."

—Professor Geoff Marcy, Center for Integrative Planetary Science, UC Berkeley

"In the past decade we have gone from complete ignorance of extra-solar planets to the verge of finding habitable planets. In *The Crowded Universe*, Alan Boss gives an extraordinary inside look at the people and events that have shaped the field. The excitement of discovery shines in Boss's elegant prose, and the work of centuries is seamlessly assembled for the non-expert reader."

—Dr. Paul Butler, Carnegie Institution of Washington

"The discovery of exoplanets has transformed modern astronomy. In *The Crowded Universe*, renowned expert Alan Boss offers an exciting insider's account of the quest for other Worlds."

—Michel Mayor, Professor of Astronomy, University of Geneva

"Alan Boss is widely respected for his scientific research and for his ability to clearly convey forefront research to the public. *The Crowded Universe* is a delightful read that chronicles the twists and turns of the birth and evolution of the rapidly evolving field of exoplanet discovery."

—Debra Fischer, Professor of Astronomy, San Francisco State University

The Crowded Universe

The Race to Find Life Beyond Earth

ALAN BOSS

BASIC
BOOKS

A Member of the Perseus Books Group
New York

Hardcover first published in 2009 by Basic Books,
A Member of the Perseus Books Group
Paperback first published in 2011 by Basic Books

Books published by Basic Books are available at special discounts for bulk purchases in the United States by corporations, institutions, and other organizations. For more information, please contact the Special Markets Department at the Perseus Books Group, 2300 Chestnut Street, Suite 200, Philadelphia, PA 19103, or call (800) 810-4145, ext. 5000, or e-mail special.markets@perseusbooks.com.

Designed by Timm Bryson
Set in 11 point Weidemann

The Library of Congress has cataloged the hardcover as follows:
Boss, Alan, 1951-
The crowded universe : the search for living planets / Alan Boss.
p. cm.
Includes index.
ISBN 978-0-465-00936-7 (alk. paper)
1. Habitable planets. 2. Extrasolar planets. 3. Exobiology. I. Title.
QB820.B676 2009
523.2'4—dc22

2008037149

Paperback ISBN: 978-0-465-02039-3

10 9 8 7 6 5 4 3 2 1

TO GEORGE WETHERILL,

the physicist who evolved into an astrobiologist

CONTENTS

ACKNOWLEDGMENTS

This book would not have been possible without the support and enthusiasm of my agent, Gabriele Pantucci, and of my editors at Basic Books, William Frucht and Lara Heimert.

Alan Boss
Washington, DC
August 4, 2008

PROLOGUE

A New Space Race

The occurrence of Earth-like planets may be a common feature of planetary systems.

—GEORGE W. WETHERILL
(*SCIENCE*, AUGUST 2, 1991)

A new space race is under way. It is not between the United States and Russia, or between the United States and the newly space-faring nations China and Japan, but between the United States and Europe. And the Europeans have a head start of more than 2 years. Whoever wins, we are on the verge of discovering how frequently Earth-like planets occur in our neighborhood of the Milky Way Galaxy.

In early 2009, the National Aeronautics and Space Administration (NASA) will launch the Kepler Mission, the first space telescope designed specifically to detect and count the number of habitable worlds orbiting stars like our Sun. In late 2006, European scientists launched a similar but smaller space telescope, CoRoT (Convection, Rotation, and Planetary Transits), designed primarily to study the physical structure of stars. However, CoRoT turns out to be quite capable of detecting Earth-like planets, giving European astronomers the chance to beat Kepler to the grand prize. Because a rocky planet with liquid water near the surface seems to be required for organisms to originate and evolve, the frequency of such Earth-like planets is perhaps the most important unknown—yet knowable—factor in any estimate of the extent to which life has proliferated in the universe.

Will Kepler and CoRoT find that such worlds are rare or commonplace? *The Crowded Universe* argues that CoRoT and Kepler will discover abundant Earths. This opinion is based on what we already know about the hundreds of planetary systems discovered to date outside our Solar System, on observations of planet-forming disks of gas and dust around young stars, and on our theoretical understanding of how planetary systems form. If this bold assertion is proved correct by Kepler and CoRoT, the implications will be staggering indeed: it will suggest that life on other worlds is not only inevitable but widespread. We will know that we cannot be alone in the universe.

CHAPTER 1

The Struggle to Find New Worlds

Do there exist many worlds, or is there but a single world? This is one of the most noble and exalted questions in the study of Nature.

—SAINT ALBERT THE GREAT (CIRCA 1200–1280)

February 6, 1995—Gordon Walker had had enough. After 12 years of painstaking observations, he and his team had found nothing. Absolutely nothing. Nada. Zip. Zilch. Now that the revised version of their final paper on the subject had been accepted for publication in the planetary science journal *Icarus*, it was time to move on to something more likely to be productive. Null results can be important to science, but they generally do not win you fame, much less a steady job, new graduate students, or the next research grant.

Walker and Bruce Campbell, his close colleague at the University of British Columbia, were true pioneers in the field of searching for planets around other stars. In the late 1970s, they had developed an ingenious technique that would enable them to discover Jupiter-mass

FIGURE 1. **Gordon A. H. Walker of the University of British Columbia, pioneer of the Doppler technique for seeking extrasolar planets. [Courtesy of Gordon Walker.]**

planets in orbit around other stars similar to the Sun. They had put their idea to good use, spending 12 years of their lives and many nights of precious telescope time on the 114-inch (3.6-meter) Canada-France-Hawaii Telescope (CFHT) on Mauna Kea, Hawaii. From 1980 to 1992, Walker, Campbell, or a member of their group at UBC had traveled to Hawaii to spend between six and twelve nights each year looking for the first hints of a planet outside the Solar System.

Even a "gas giant" planet such as Jupiter, with 318 times the mass of Earth and 11 times its diameter, is nearly impossible to find when it is in orbit around a star. The problem is not so much the faintness of the planet itself—Jupiter is no fainter than the distant galaxies that have been imaged by the Hubble Space Telescope in the Deep and Ultra Deep Field surveys. The problem is that when a planet is orbiting a far brighter object, it is exceedingly difficult to see the planet in the glare of the star's light. Stars such as the Sun give out most of their light at visible wavelengths, where our eyes are best suited to seeing. At visible wavelengths, Jupiter is about a billion times fainter than the star it circles. Even with its incredibly powerful cameras, NASA's Hubble is incapable of snapping a photograph of a planet orbiting a star—the star's light would drown out the planet's light many times over.

Campbell and Walker developed a completely different technique for detecting the presence of a seemingly invisible planet lurking in the glare of its star. Rather than trying to see the planet directly, they would infer its presence by the effects that it must have on its own star. Their scheme relied on the fact that something we are all taught in elementary school is not quite right—the planets of the Solar System do not orbit the Sun. Rather, all the planets and the Sun itself orbit a single point in space, the center of mass of the Solar System. This center of mass is the place where the entire Solar System could be balanced on a fulcrum if there were a teeter-totter large enough for all to join in. To balance a teeter-totter, an adult must sit much closer to the central fulcrum than a small child sitting on the other end of the teeter-totter. Similarly, because Jupiter is 1000 times less massive than the Sun, the balance point for the Jupiter–Sun system is 1000 times closer to the Sun than to Jupiter. As Jupiter orbits the Sun over a period of 12 years, the Sun orbits the center of mass, or barycenter, of the Solar System in a circle that is 1000 times smaller than Jupiter's orbit. If astronomers living on a planet around a nearby star were willing to spend 12 years watching the Sun, they might be able to detect this periodic motion of the Sun. If they did, the only explanation possible would be that the Sun must be orbited by a planet 1000 times less massive. There is no other physical explanation for a star appearing to wobble back and forth across the sky this way.

Because of the presence of Jupiter, the Sun moves around the center of mass of the Jupiter–Sun system on a circle whose diameter is roughly equal to the diameter of the Sun itself. Detecting such a miniscule wobble of the Sun from the great distance of another star is not easy, but compared to trying to take a direct photograph of an extrasolar Jupiter, it is simple. Walker and Campbell decided to hunt for Jupiters by clocking the speed of stars as they move around the barycenters of their planetary systems. The Sun moves with an orbital speed of about 30 miles per hour, or 13 meters per second, around the Sun–Jupiter barycenter. A speed of 30 mph sounds like it

FIGURE 2. Christian Johann Doppler [1803–1853], the Austrian physicist who showed that sound waves shift in wavelength by an amount that depends on the velocity of their source. [Courtesy of Wikimedia Commons.]

should not be hard to detect; if you are traveling 30 mph over the speed limit, you can be sure that a police officer will have no trouble pointing a radar gun at you and writing you a speeding ticket that will stand up in court. The problem for stars is that the relevant standard of comparison is the speed of light, which is about 186,000 miles per second, or 670,000,000 mph. Thus 30 mph is practically stationary by comparison.

The star's speed is compared this way because Walker and Campbell planned to find planets by measuring it through the Doppler effect. Christian Johann Doppler, an Austrian physicist, hypothesized in 1842 that light waves emitted by a moving star behave exactly the same way as sound waves emitted by a moving train. When a train blows its whistle as it is moving toward you, you hear the whistle at a higher pitch, or frequency, than when the whistle is sounded while the train is stopped. When the train is traveling away from you, the whistle sounds lower in pitch than when it is at rest. The distinctive change in pitch of the sound of an automobile engine as it passes by is familiar to NASCAR fans, who are dazed by the roar of the Doppler effects created by dozens of stock cars racing by at speeds of 180 mph or more.

The change in frequency of the sound produced by the Doppler effect depends on the ratio of the speed of the NASCAR race car to the speed of sound. The sound speed in air is about 750 mph, or 340 meters per second, so a race car can move at about one-quarter of the speed of sound. That means that the sounds heard during a NASCAR race can be shifted to higher or lower pitches by a similar fraction—

a distinctive change equivalent to the difference in pitch between the musical notes A and D.

Doppler postulated that this frequency shift would occur for light in the same way it works for sound, because light and sound are both wave phenomena. He hoped the effect would explain the red and blue colors of different binary stars, systems where two stars are in orbit about their own center of mass, with red stars moving away from the observer, and hence emitting lower-frequency and longer-wavelength light, and blue stars moving toward the observer, producing light of higher frequency and shorter wavelength. Sadly, he was mistaken, because the frequency shift associated with the Doppler shift of light is small indeed compared to the intrinsic variations in the colors of stars, whether they are in binary systems or all on their own. Even binary stars moving with speeds of several miles per second exhibit a frequency shift of their light of only one part in 100,000. The French physicist Armand Hippolyte Louis Fizeau (1819–1896) independently predicted the Doppler effect for light in 1848, leading the French to describe it as the Doppler-Fizeau effect.

The zenith of Doppler's career came in 1850, when he was appointed founding director of the Institute of Physics at Vienna's Imperial University. As director, he was responsible for deciding which candidates would be admitted for study at the university. (One candidate he turned down was Johann Gregor Mendel, who later gained admission to the university through a different department and became a pioneer of genetics research. Mendel's mathematical skills were not considered on a par with what Doppler required of a physicist, to the everlasting benefit of modern genetics.)

The Doppler effect that Walker and Campbell sought to measure was quite small. If the motion of the Sun induced by the presence of Jupiter is 30 mph, dividing that speed by the speed of light—670,000,000 mph—yields a Doppler shift of about one part in

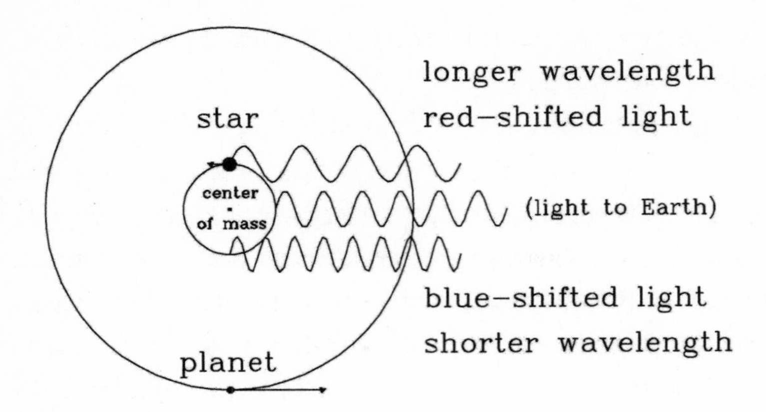

FIGURE 3. The presence of a planet can be inferred from the Doppler shift of light emitted by the star as it orbits the center of mass of the star–planet system.

20,000,000. Campbell and Walker devised a means of measuring such a small shift by including a glass-ended container filled with hydrogen fluoride gas in their telescopic instrumentation. The hydrogen fluoride was introduced to serve as a stable set of reference lines for measuring the expected tiny Doppler shift. The light from the target star would pass through the telescope, bounce off the telescope's mirrors along the way, and then pass through the hydrogen fluoride cell before it could enter the telescope's spectrograph. There, the star's light would be split into different colors, just as a prism splits white light into the colors of the rainbow. The hydrogen fluoride gas absorbs some of the star's light only for very specific, narrow bands of color (wavelengths), thereby superimposing a set of absorption lines for comparison with the emission and absorption lines in the star's spectrum. The star's spectral lines are produced by the ions of elements, such as calcium, sodium, and iron, and by molecules, such as titanium oxide, in the star's outer atmosphere.

As the target star moves in orbit around the barycenter, alternating between moving toward Earth and moving away from Earth, the star's spectral lines will vary in frequency (equivalently, in wavelength or color) according to the Doppler effect. Meanwhile, the hydrogen fluoride lines do not change in frequency at all, so they provide a stable reference for making precise measurements of the changes in frequency of the star's spectral lines. Given that it would take 12 years of observations to follow the stellar wobble induced by a planet similar to Jupiter, the hydrogen fluoride cell provided the stable reference source that is crucial to carrying out such a long-term search. Hydrogen fluoride had the particular advantage of providing a number of widely spaced lines with just the right colors to optimize the search, but it also had the disadvantage of being poisonous. Accidentally breaking the cell that contained the hydrogen fluoride would halt the observing program—and possibly an astronomer or two. With the hydrogen fluoride cell in place, Walker was able to make Doppler measurements accurate to about 35 mph, or 15 meters per second, close enough to the expected wobble of 30 mph induced by a Jupiter-like planet to begin the search.

Walker's planet search was granted a special status on the Mauna Kea telescope as the only long-term program during the years 1980–1992. This meant that Walker and his group did not have to worry about producing new discoveries every year in order to be awarded their allotment of telescope time for the following year, as is usually the case for astronomers. (Most telescopes are oversubscribed, so the committees charged with assigning telescope time often have to be brutal and reject nonproductive ongoing projects.) Walker had the luxury of being able to search for hidden beasts that could be detected only by taking Doppler measurements for a decade or longer: gas giant planets similar in mass to Jupiter. With their hydrogen fluoride cell accuracy of 35 mph, in fact, Walker could hope to find planets only of Jupiter mass and above; there was no hope of searching for the much smaller Doppler shift that would be produced by an

Earth-like planet. Still, judging on the basis of the only planetary system known at the time, our own, Jupiters appeared to be natural products of the planet formation process, so they should be out there, waiting to be discovered. A Jupiter-mass planet on a Jupiter-like (12-year) orbit around a Sun-like star would be expected to be a signpost for Earth-like planets in the same planetary system.

Bruce Campbell thought that they had found something earlier. In 1988 he and Walker published a paper in the *Astrophysical Journal* reporting their results to date. Of the 16 stars on their target list at that time, 7 showed long-term trends that were consistent with the wobbles induced by planets with masses in the range of 1 to 9 Jupiter-masses. They suggested that they had found "the tip of the planetary mass spectrum." There seemed to be good evidence for a 1.7-Jupiter-mass planet on an orbit with a period of 2.7 years around one of the stars in a binary star system called Gamma Cephei, the third brightest star in the constellation Cepheus.

In 1992, however, Walker published a paper in the *Astrophysical Journal* that retracted the 1988 claim for a planet in the Gamma Cephei system. By then, Bruce Campbell had quit the planet search program and left astronomy. Four more years of data had changed the orbital period of the suspected planet from 2.7 years to 2.5 years, the same period at which they found that the star's calcium emission line varied. The variation was thought to be caused by the rotation of a giant star in the Gamma Cephei system with a rotational period of 2.5 years. A rotating star can also produce a Doppler shift if the surface of the star is covered with spots and other irregularities that affect the amount of light coming from different regions, some of which are moving away from Earth, and some toward it. Gamma Cephei's "planet" seemed to be nothing more than grumblings from a cantankerous old star.

The paper that Gordon Walker and his remaining colleagues published in 1995 stated that they had looked for Doppler wobbles in 21 stars over a period of 12 years and had found no firm evidence for any planets with masses greater than Jupiter on Jupiter-like orbits. Considering this outcome along with the results of other planet searches, they furthermore claimed that 45 nearby stars showed no evidence of Jupiter-like planets and stated that their results presented a "challenge to theories of planet formation."

July 27, 1995—George Wetherill was pleased to have his own revised paper accepted for publication in *Icarus*. Even for a distinguished, prize-winning senior scientist like Wetherill, former director of the Carnegie Institution of Washington's Department of Terrestrial Magnetism, it was a relief. The review process for a scientific paper can be a long and painful one, wherein the often anonymous reviewers lob grenades in the general direction of the author, while the journal's editor plays the role of a neutral bystander, trying to avoid the shrapnel. Wetherill got off relatively easily; he had to suffer only five months of anxiety before his revisions were considered acceptable to the reviewers and hence to the editor.

FIGURE 4. George W. Wetherill [1925–2006], the pioneer of modern theoretical work on the formation of habitable planets. [Courtesy of Janice Dunlap (Carnegie Institution).]

Wetherill had studied what would happen to the formation process of habitable planets if he varied some of the basic parameters, such as the mass of the central star, the mass of the planet-forming disk of gas and dust, and the effects of the giant planets.

The calculations he ran were based on the Monte Carlo technique, an apt name for what was going on. The route through which rocky planets such as Earth are built up from a large population of smaller bodies is a random one.

This lengthy process entails the banging together of progressively larger and larger solid bodies as they orbit their central star. It is thought to begin with tiny dust grains too small to be seen by the human eye. Wetherill had taken a shortcut and was studying the final phase of the collisional accumulation process, when a swarm of several hundred lunar-mass bodies are in orbit, having taken about 100,000 years to grow that large. This final phase takes tens of millions of years to play out, because the lunar-sized bodies have to wait longer and longer to smack into each other and grow even larger. The Moon's mass is 81 times smaller than that of Earth, so hundreds of fierce collisions between sizable bodies are needed to form a planet as large as Earth.

The lunar-sized bodies are spread out over an immense area, and the chances of random collisions occurring are small. The situation is like a Dodge-Em car ride at a State Fair, but one in which dozens of cars are allowed to roam over the entire state of Kansas, instead of being confined to an area the size of a basketball court. In the planet formation ride, when two bodies of equal mass collide, they are likely to stick together and form a new body with twice the mass, instead of just bouncing off each other. Also, lunar-mass bodies are large enough that their gravitational attractions for each other, though small, are strong enough to determine the ultimate outcome of the ride. Mutual gravitational interactions pull the careening bodies toward each other, resulting in near misses, sideswipes, head-on collisions, and rear-end impacts without the benefit of airbags.

The entire process is an unplanned, chaotic one, and changing the initial position of just one of the Dodge-Em cars at the start of the ride can result in planets too close or too far from the Sun to be habitable. As a result, Wetherill had to run many different versions of the same basic simulation of the rocky planet formation process in order to get a good statistical picture of what the most likely outcomes would be when he played with the various free parameters in the modeling effort. For the *Icarus* paper, Wetherill ran about 500 Monte Carlo models in order to study 20 different assumptions about the mass of the star, the mass of the planet-forming disk of gas and dust, and the effects of the giant planets.

What he found was tremendously reassuring for those who imagined that planetary systems like our own might be common. Essentially all of the variations he tried resulted at times in one or two habitable worlds—that is, roughly Earth-mass bodies orbiting at distances from their stars where it would be warm enough to have liquid water on their surfaces, but not so hot as to turn the water into steam. Stars with masses half as large as the Sun or half again as large as the Sun appeared to be likely to have Earth-like planets. And for stars with the same mass as the Sun, the formation of at least one habitable planet seemed inevitable. When the amount of mass in the disk was doubled or halved, habitable worlds still formed, but with twice the mass each or half the mass each, respectively.

Wetherill noted that the presence and location of the gas giant planets that were assumed already to have formed in these systems by the time he turned on the electricity for the Dodge-Em cars were crucial for determining the final outcome of the ride. The gravitational forces exerted by the gas giants determined how far out from the central star rocky planets could grow, just as they did in our Solar System. Were it not for the gravitational yanks from Jupiter and Saturn, Mars would probably be a larger planet than it is today, with about one-tenth the mass of Earth, and there would be another rocky planet beyond Mars, instead of the horde of frustrated "wannabe

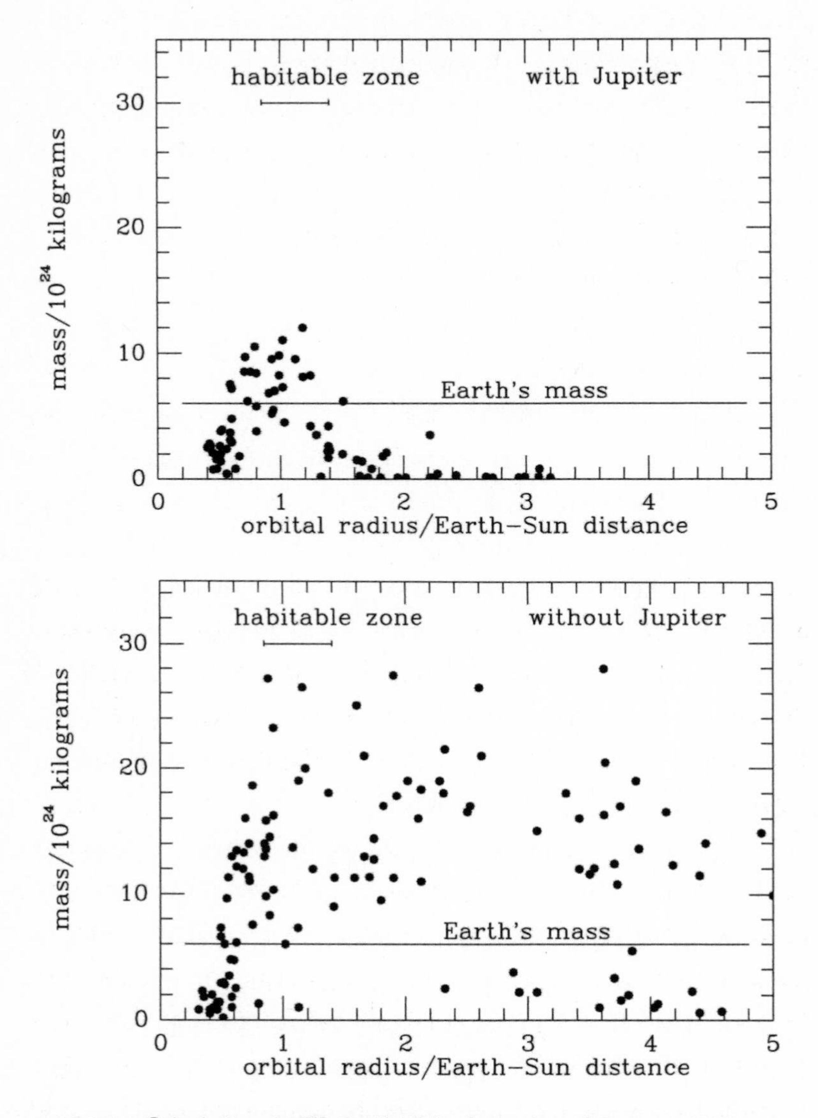

FIGURE 5. Calculations by Wetherill of the formation of the terrestrial planets through impacts between lunar-mass planetary embryos, showing a variety of possible outcomes depending strongly on the presence or absence of Jupiter. [Adapted from G. W. Wetherill, 1996, *Icarus*, volume 119, pages 226 and 235.]

planets" in the asteroid belt. When Wetherill ran models without any gas giant planets at all, he found that the planets that formed were likely to be twice as massive as when Jupiter and Saturn were up to their dirty tricks, and that the planets extended all the way out to Jupiter's orbital distance, five times Earth's distance from the Sun (93,000,000 miles, or 150,000,000 kilometers, a distance that is called the astronomical unit, or AU).

Wetherill concluded that "abundant populations of habitable planets" could be produced for stars of any of the masses he considered. On purely theoretical grounds, Wetherill had shown that one could expect that habitable planets were abundant in the universe.

August 21, 1995—The August issue of *Icarus* arrived on schedule in the Department of Terrestrial Magnetism library, carrying with it the news of the failed planet search of Gordon Walker and his colleagues. Walker's paper, with the null results of their 12-year search for "Jupiters," was unnerving. Where were the Jupiters we were all expecting to find in great abundance?

Earlier in the year, the January 20 issue of *Science* had included a paper in which I discussed the chances for detecting the first Jupiter-mass planets around nearby stars. I had argued that, on the basis of computer models I had calculated of how planet-forming disks would heat and cool, it was to be expected that gas giant planets would form at Jupiter-like distances even around stars with lower masses than the Sun. Such "red dwarf" stars are much more common in the Sun's neighborhood of the Galaxy than are stars like the Sun or more massive than the Sun. Because of their relative closeness and abundance, red dwarf stars seemed like the first place to look for another planetary system similar to our own. My calculations implied that if red dwarfs had gas giant planets, they would be orbiting far enough away from their stars for there to be rocky terrestrial planets orbiting comfortably in the habitable zone of the star. Red dwarf stars, the most common type of star, could thus be home to the majority of the habitable planets in the Galaxy.

Gordon Walker's search had concentrated primarily on stars like the Sun, rather than on the more numerous and nearby red dwarfs. We had all assumed that Sun-like stars would have Jupiter-like planets, based on the undeniable evidence of the one obvious example we had—our own Solar System. But maybe the Solar System was not the proper example to consider after all. George Wetherill had suggested as much several years before, during a talk on finding new planetary systems that he had delivered at a conference held at Caltech, in Pasadena, California, on December 7–10, 1992. Wetherill had shocked the audience of several hundred astronomers and planetary scientists by pointing out that the mere fact that the one known planetary system contained a Jupiter did not necessarily mean that Jupiters were common. Jupiter protects us from the comets that revolve in the Kuiper Belt beyond Neptune's orbit. When a malevolent comet decides to break out of the Kuiper Belt and make a suicidal dash toward Earth, Jupiter plays the role of the batsman protecting the wicket in a cricket match. It swats the comet out of the Solar System, or forces it to smash harmlessly into the Sun, or takes it right in the face, as Jupiter did with the startling collision of Comet Shoemaker-Levy 9 just 2 years later, in 1994.

Without Jupiter, Wetherill noted, Earth would be whacked by comets roughly 1000 times more often than is the case with Jupiter at bat. Dinosaur-killing events such as the collision that marked the Cretaceous–Tertiary extinction event 65 million years ago would be occurring every 100,000 years, instead of every 100 million years or so. It is hard to imagine intelligent life evolving if the evolutionary clock were reset by that kind of catastrophe every 100,000 years. Wetherill made the point that we therefore did not have a single unbiased example of a planetary system to use to predict what would be found elsewhere in the Galaxy. The problem of interpreting the statistics of a sample composed of only a single solar system was eliminated, because even that sample was biased and had to be rejected. We really had not a single planetary system to use to predict how common Jupiters would be.

Wetherill's 1992 warning combined with Walker's null results to yield the depressing realization that we might have been misled—Jupiters might be the exception rather than the rule. How were we going to find the first extrasolar planet if Jupiters were that rare? Would we be faced with searching for "failed Jupiters"—planets, such as the ice giants Uranus and Neptune, that failed to gather as much gas from the planet-forming disk as Jupiter and Saturn did and, accordingly, ended up with masses 10 times smaller than the gas giants? The task of finding an extrasolar planet by the indirect Doppler search method would be 10 times harder, because the star would be sitting 10 times closer to the fulcrum on the teeter-totter, making its Doppler wobble 10 times smaller.

Alexander Wolszczan of the Arecibo Observatory in Puerto Rico had used the observatory's giant radio telescope in 1992 to find a couple of several-Earth-mass bodies in orbit around a dead star, a pulsar, through a variation of the Doppler timing method. However, there is no beach-front property on a pulsar planet—no reason to point an antenna in its direction in the hope of picking up radio transmissions playing the songs on an extraterrestrial Top Ten list. The pulsar planets are likely to be parched cinders immersed in a deadly rain of radiation from the pulsar, a rain that would make the fallout from a nuclear bomb look harmless by comparison. The universe was looking to be devoid of life.

October 6, 1995—A surprising discovery was announced at a conference being held in Florence, Italy. Michel Mayor, a Swiss astronomer from the Geneva Observatory, had decided to join the search for planets after spending several decades looking for unseen binary star companions by the Doppler technique. The year before, he had installed a new spectrograph on the 77-inch (1.93-meter) telescope of the Haute Provence Observatory, near Marseille, France, and this spectrograph was capable of making Doppler measurements accurate to about 30 mph (13 meters per second), similar to what Gordon Walker had achieved in Hawaii. Mayor did not adopt the potentially dangerous hydrogen fluoride gas cell that Walker had developed; he relied instead on the light emitted by a lamp containing thorium and argon, two relatively nonpoisonous gases, for his wavelength standard.

Mayor had a list of 142 stars similar to the Sun that did not seem to be members of binary star systems, and he intended to see whether he could find any very-low-mass companions to these stars, and perhaps a gas giant planet or two. Mayor began the search in April 1994 on the 77-inch telescope. In September 1994, Mayor and his junior colleague, Didier Queloz, began observing a solar-type star called 51 Pegasi in the constellation Pegasus. By January 1995, they had taken enough data on 51 Pegasi to realize that something was making the star bounce around, but by then Earth's orbital motion around the Sun had put the constellation in the direction of the Sun, making it difficult to observe. Mayor and Queloz had to wait for Pegasus to return to the night sky in July 1995, and once it did, they banged away at 51 Pegasi for eight nights in a row. Once they finished that marathon observing run and analyzed their data one more time, they realized that they had bagged a planet. But just to be sure, they waited two more months until September and took even more data on 51 Pegasi. Then Mayor and Queloz quickly wrote a short paper and submitted it to the journal *Nature.*

While their 51 Pegasi discovery paper was still in the process of being scrutinized for publication, Mayor decided to break the press embargo imposed by *Nature* and announce their finding at the Florence meeting. Mayor showed their data for 51 Pegasi, which showed the Doppler wobble of the star varying by more than 120 mph, or 55 meters per second, a variation significantly larger than their errors. With numerous repeated observations taken over two observing seasons, it was clear that they had something. The data could be fit by the wobble predicted if 51 Pegasi had a planet with a mass of at least half the mass of Jupiter moving on a circular orbit.

That all sounded good. The only odd thing was that 51 Pegasi's planet had an orbital period not of 12 years, like Jupiter, but of just 4.2 days. This meant that according to Kepler's Third Law of planetary motion, 51 Pegasi's planet orbited its star at a distance that was 100 times closer than Jupiter—that is, at a distance 0.05 times the distance between Earth and the Sun. Such a planet would be in danger of being boiled away so close to its star. Indeed, 51 Pegasi's planet would have to be nearly as hot as some stars, with an atmosphere heated to thousands of degrees Fahrenheit, hot enough to be slowly losing mass as the gas molecules escaped into space. This new world was a "hot Jupiter" very unlike the frosty conditions in the atmosphere of our Jupiter, where the temperature is hundreds of degrees Fahrenheit below zero.

Two days later, the *Washington Post* reported this claim for the discovery of an extrasolar planet on page 36, mistakenly attributing it to Italian rather than Swiss astronomers.

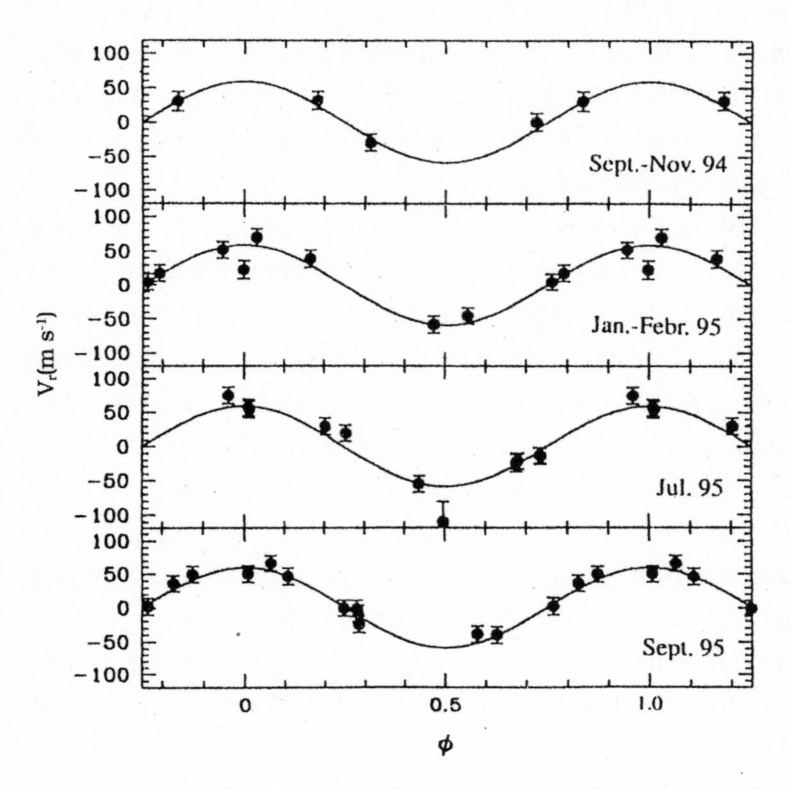

Figure 6. Discovery of the first extrasolar planet orbiting a Sun-like star, showing periodic changes in the radial velocity of the star 51 Pegasi measured through its Doppler shift. [Reprinted, by permission, from M. Mayor and D. Queloz, 1995, *Nature*, volume 378, page 356. Copyright 1995 by Macmillan Magazines Limited.]

The word was out about 51 Pegasi's planet. Given the amazingly short orbital period, it would not be hard for another group to try to replicate the Swiss results. Pegasus was still up in the night sky, and anybody with a good spectrometer and four nights of telescope time could see whether the Swiss were right or not. San Francisco State University astronomer Geoffrey Marcy and R. Paul Butler of the University of California, Berkeley, had their own planet search under way at the Lick Observatory's 120-inch telescope on Mount Hamilton, just east of San Jose, California. Marcy and Butler spent parts of four nights of 120-inch time to follow 51 Pegasi's every wobble, using a cell filled with iodine gas as their reference standard. They found that Mayor and Queloz had been right on the money: 51 Pegasi had a half-Jupiter-mass planet on an outrageously short-period orbit. They reported their confirmation of the Swiss breakthrough, and this time the *Washington Post* carried the story on the front page of the October 19 issue, less than two weeks after the Florence announcement. We finally had a real extrasolar planet around a Sun-like star, and it was a Jupiter to boot, albeit a hot Jupiter.

November 23, 1995—Mayor and Queloz's epochal paper detailing the discovery of 51 Pegasi's planet appeared in *Nature*. In an accompanying commentary, Gordon Walker summarized the woeful history of the field to date, including his own efforts. He concluded by noting that the short orbital period would be a challenge to theorists and wondering, "How did it get there in the first place?" Yet the reasons why Walker did not find such a hot Jupiter were clear in retrospect. Mayor and Queloz had searched 142 stars to find a single hot Jupiter, whereas Walker and his group had followed only 21 stars, a sample too small for there to have been a good chance of detecting a phenomenon that may occur only about 1% of the time. Furthermore, 51 Pegasi's planet had the entirely unexpected orbital period of 4.2 days, and Walker did not search his data for such a short-period orbit. The shortest orbital period he considered was 40 days, which must have seemed ridiculously short at the time.

51 Pegasi b, the astronomical name for the first planet orbiting a Sun-like star (it is nicknamed 51 Peg b, which does not really help), was real. The Swiss astronomers had proved that the Doppler

technique worked, in spite of the inability of the Canadian team to find a planet after they invented the basic calibration technique. The world's other Doppler planet search teams decided it was high time to stop looking only for 12-year periodicities in their data and to start looking for some more hot, or at least warm, Jupiters, with shorter-period orbits.

The hunt was on. There were planets to be found in the hills and mountains, where small groups of telescopes waited patiently for sunset.

CHAPTER 2

Eccentric Planets

The Sun does not move.

—LEONARDO DA VINCI (1452–1519)

And yet it does move.

—GALILEO GALILEI (1564–1642)

January 16, 1996—Geoff Marcy and Paul Butler had some hot new stuff to present at the American Astronomical Society meeting being held in San Antonio, Texas. A press conference was scheduled as the first event of the following day at the San Antonio Hilton Hotel. Marcy's talk was entitled simply "New Planets."

Marcy and Butler had been running their own Doppler planet search for almost a decade and had been monitoring the movements of 120 stars similar to the Sun. Following their confirmation of 51 Peg's planet in October, they had spent the months of November and December cranking through the Doppler data that they had been assiduously gathering for 8 years. A shortage of the necessary computer power had hampered their ability to analyze the data. Marcy and Butler had been running their planet search program on a shoestring budget the entire time. After the discovery of 51 Pegasi b, they were able to secure the use of six new Sun workstations. Butler spent most of his time before the San Antonio meeting putting the workstations to good use analyzing their data, searching for hidden periodicities in the Doppler wobbles that would signal the possibility that another giant planet stallion was galloping around one of the stars.

By the time January rolled around, Butler had been able to analyze the data for only half of their 120 targets. But his efforts were not in vain. By early December, they already knew that they had won the race for second and third place in the Doppler Sweepstakes, when two of Butler's targets crossed the finish line. Marcy and Butler decided to keep their discoveries a secret until they had the chance to make their announcement in San Antonio.

January 17, 1996—Marcy was scheduled as the last speaker in the American Astronomical Society meeting press conference, so the astronomers, reporters, and television crews with their bright lights on and cameras running would all have to sit and wait until the very end to find out what had been found. Even though the press conference had been moved to a larger room than originally planned, the new room had filled up fast and stayed filled, with late-arrivers forced to stand up next to the walls.

Marcy did not disappoint. He led off with the confirmation of the reality of 51 Pegasi b and then announced the surprise of the day: he and Butler had tripled the number of known extrasolar planets by discovering two new planets around the stars 47 Ursa Majoris and 70 Virginis. Both of these discoveries were around solar-like stars, located in the constellations of the Big Dipper and Virgo, respectively, no more than 80 light-years away (a light-year is the distance that light travels in 1 year; it is equal to about 6 trillion miles, or about 10 trillion kilometers). 47 Ursa Majoris's planet had a mass no less than 2.4 Jupiter masses, and it completed a circular orbit of the star every 3 years. This implied an orbital distance about twice that of Earth from the Sun—less than half that of Jupiter, but 40 times farther out than 51 Pegasi's hot Jupiter. Evidently 47 Ursa Majoris had a cool Jupiter like the Sun, a planet that was presumably a gas giant, just as we had all been expecting and hoping to find.

Marcy could offer only lower limits on the mass of 47 Ursa Majoris b, because the Doppler effect is sensitive only to motions directly along the line of sight between the telescope and the target star. If the star is wobbling like mad back and forth across the sky, rather than to and fro along the line of sight, the Doppler shift will be zero. Thus if a star happens to be orbited by a planet whose orbital plane is perpendicular to the line of sight, the planet cannot be detected by the Doppler effect. If the planet's orbit were tilted such that there was just a little to-and-fro motion, Doppler spectroscopy could detect that motion but would assign a mass to the planet much lower than the true mass, because it would be sensing only a small fraction of the entire range of changes in speed of the star around the system's center of mass. Hence 47 Ursa Majoris's planet could be considerably more massive than 2.4 Jupiter masses.

The chances are small that a planet's orbital plane will be aligned in such a way as to be perpendicular to the line of sight, so the minimum mass found by the Doppler technique generally is not a bad estimate. For a random distribution of planetary orbital planes in the sky, the average correction is only a factor of 1.3, meaning that the chances were that the true mass of 47 Ursa Majoris's planet was not 2.4 Jupiter masses, but about 3.1 Jupiter masses.

Marcy next announced 70 Virginis's planet, with a mass of at least 6.6 Jupiter masses, orbiting every 117 days at a distance just greater than Mercury's orbit. At that distance, this behemoth was a warm Jupiter. Mercury is strongly heated by the radiation from the Sun, which is seven times stronger at Mercury's distance than at that of Earth's orbit. We now had a family of planets reminiscent of the Three Bears, with a hot Jupiter, a warm Jupiter, and a cool Jupiter. The cool Jupiter was far enough out from its star, 47 Ursa Majoris, that there might be room between the two for an Earth-like world to orbit.

Perhaps more surprising than its large mass was the fact that 70 Virginis b was on an eccentric orbit, an elliptical orbit unlike the more nearly circular orbits of Earth and Jupiter. On the eccentricity scale of 0 to 1, where a circular orbit has an eccentricity of 0, Earth has an orbital eccentricity of 0.02, Jupiter of 0.05, and 70 Virginis b of 0.4.We knew that binary stars typically have eccentric orbits, whereas the major planets in the Solar System do not. Binary stars form from the collapse of dense clouds of gas and dust, which can break up into two or more fragments during the free-fall collapse phase. This process is rapid and somewhat unpredictable, like removing an unwanted tall building by exploding dynamite charges around its foundations. Collapsing gas clouds tend to form young protostars that are on highly noncircular, elliptical orbits as a result.

Planets, though subject to their own highly chaotic formation processes, were expected to end up on nearly circular orbits, because the net result of all of the wild collisions needed to form a rocky planet such as Earth, whacking Earth first one way and then the other, was thought to average out to a roughly circular orbit. This was the result suggested by planet formation calculations like those of George Wetherill and others, and it was what characterized the

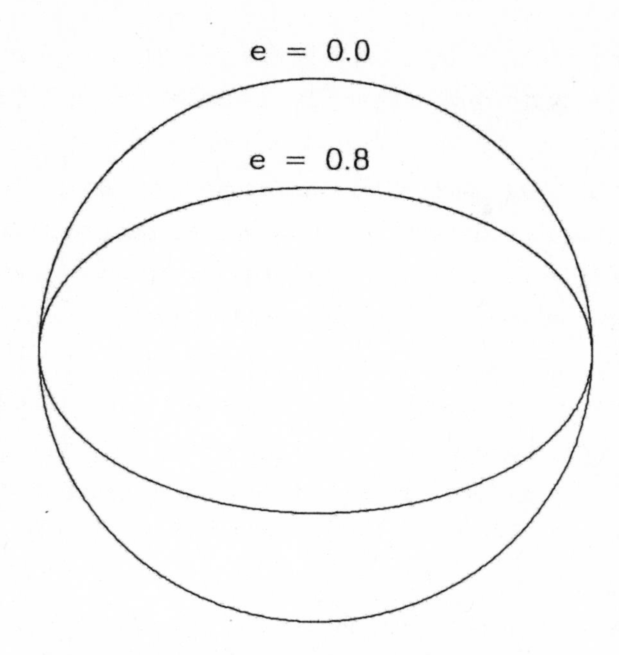

FIGURE 7. **Circular orbits have zero eccentricity, whereas eccentric orbits with eccentricities between 0 and 1 are elliptical in shape.**

major planets of our Solar System, so no one had expected to find a highly eccentric planet like that around 70 Virginis. Maybe 70 Virginis b was just an oddball, and eccentric planets were indeed rare. Or maybe these massive planets were not really "planets" at all, but merely low-mass objects that had formed in the same way that binary stars form, and so had ended up on eccentric orbits.

Marcy and Butler still had 60 more stars to analyze.

June 12, 1996—Robert Brown of the Space Telescope Science Institute in Baltimore asked me to join a new NASA committee that would help develop the science justification for the Space Interferometry

Mission (SIM). The SIM Science Working Group would have a subcommittee that would examine the ability of SIM to find planets. Given that I had been advising NASA about its plans to find planets around other stars since 1988, it was an offer I could not resist.

SIM was a brand new acronym for an idea that had been around for over a decade: a plan for a major space telescope. SIM was intended to be the first interferometer in space working at visible wavelengths. The resolving power of a telescope is limited by two things, the diameter of the telescope and the wavelengths of the light it sees. Sharper images require higher resolving power. Although the Hubble Space Telescope has a primary mirror with a diameter of 94 inches (2.4 meters), affording the sharpest views to date of the heavens from its perch well above the blurring produced by Earth's churning atmosphere, astronomers would love to have a space telescope 10 times larger that can see things 10 times smaller yet. Given that Hubble cost several billion dollars, though, and that the cost of anything in space tends to scale directly with its mass, a telescope 10 times larger might well cost 100 to 1000 times as much, if it were designed to operate in the same way as the Hubble telescope. Clearly, such a telescope was not going to be built.

The interferometer, however, would get the resolving power of a much larger telescope by combining the light from a number of smaller telescopes separated by a distance equal to the diameter of the desired larger telescope. SIM would string together seven small telescopes along a boom 33 feet (10 meters) long and combine the light of the seven telescopes into a single interferometer with the unprecedented resolving power of a space telescope effectively 33 feet in diameter.

Not only would SIM be able to take photos four times sharper than the Hubble telescope, but it would also be able to detect planets by measuring the sideways wobbles of their stars around the centers of mass. Instead of measuring the Doppler change in speed of the star as it orbited the barycenter, SIM would watch as the star moved across the sky on a tiny orbit no larger that its own diameter. By comparing a star to distant background stars, which have no perceptible motion on the sky, SIM would then be able to see which nearby stars were wobbling back and forth, thus indicating unseen planets.

SIM would use the indirect planet detection technique called astrometry, which had been pioneered with a ground-based telescope at Swarthmore College's Sproul Observatory by Peter van de Kamp. Van de Kamp had become director of Sproul in 1937, and the following year he began an astrometric planet search effort that concentrated on Barnard's star, a red dwarf star with one-seventh the mass of the Sun. Low-mass stars make good targets for astrometric planet detections, because such stars are easier for their planets to move around the center of mass. Barnard's star must sit farther out on the teeter-totter to balance a given planet, making its wobble seven times larger than that of the Sun. Barnard's star had been discovered in 1916 by Edward Emerson Barnard, then at Chicago's Yerkes Observatory, who found that it was one of the closest stars to Earth, only 6 light-years away (only the Alpha Centauri triple star system, at a distance of about 4 light-years, is closer). Close is good for astrometry, because the astrometric wobble becomes 10 times smaller, and so 10 times harder to see, when the star is 10 times farther away. In spite of its being so close, the fact that Barnard's star is a faint red dwarf means that it cannot be seen by eye, but it was a piece of cake to photograph with van de Kamp's 24-inch-diameter telescope at Sproul.

Van de Kamp followed Barnard's star religiously for 25 years before deciding that he had enough evidence for a planet. In 1963 he

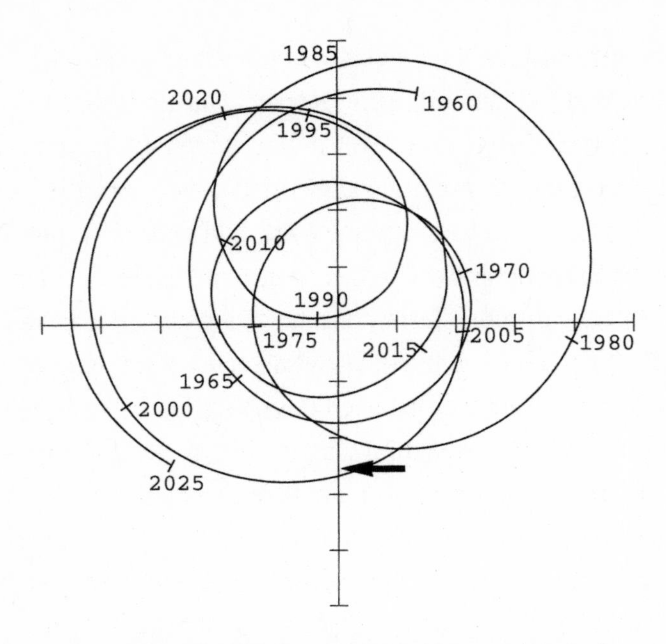

FIGURE 8. Motion of the Sun around the center of mass of the Solar System from 1960 to 2025. The Sun wobbles back and forth over a distance roughly twice its own radius [black arrow]. [Courtesy of Robert A. Brown (Space Telescope Institute) and Michael Shao (Jet Propulsion Laboratory). Based on JPL's DE200 planetary ephemeris.]

published a paper in the *Astronomical Journal* showing the results: Barnard's star seemed to wobble in a manner suggestive of its having a planet with a mass of 1.6 Jupiter masses, orbiting the red dwarf at a distance of 4.4 AU, similar to Jupiter's orbital distance of 5.2 AU. The orbit was oddly eccentric and so was troubling, but otherwise it was what astronomers had been expecting for the first extrasolar planet. Barnard's star was duly entered into the textbooks as the first example of a star with a giant planet.

Some other astrometrists, however, were not convinced of the reality of Barnard's star's planet. For one thing, several previous claims of astrometric planet detections had since been discarded or disproved. George Gatewood, a graduate student at the University of Pittsburgh, was strongly encouraged by one of his professors to reexamine the evidence for the planet. Gatewood at first refused, but after the professor repeated the suggestion several more times, Gatewood relented and started work on Barnard's star. He did not use the 2400-odd photographic plates that van de Kamp had gathered over the decades but, rather, examined a smaller collection of 241 glass plates that had been taken at Pittsburgh's Allegheny Observatory and at Wesleyan University's Van Vleck Observatory. Instead of making the measurements by hand with a mechanical plate-measuring machine, as van de Kamp and his students had done at Swarthmore, Gatewood analyzed the positions of Barnard's star with respect to the background stars with a new automatic plate-measuring machine that had been devised by the astronomers at the U.S. Naval Observatory, in Washington, D.C. Gatewood finished his doctoral research on Barnard's star in 1972 and published the results in the *Astronomical Journal* the following year. His analysis found no support for a planet around Barnard's star. Van de Kamp had been fooled by several subtle changes to the Sproul telescope, changes accompanying the installation of a new cell to hold a lens, by the use of a different type of photographic emulsion for the glass plates, and by a lens focus adjustment. In the succeeding years, van de Kamp's claim slowly disappeared from the textbooks. Van de Kamp passed away in 1995, before the announcement of 51 Pegasi's planet, but he continued to believe that Barnard's star had a planet, even if no one else did.

Astrometric planet detection acquired a reputation as a dubious enterprise, rather like the search for life on Mars, which was associated in astronomers' minds with the claims for Martian "canals" that must be signs of an intelligent civilization on our neighboring planet. Proper astronomers did not stoop to looking for planets or searching for life in the Solar System.

Given this sordid history, it would take someone with guts to propose a major space telescope intended to, among other things, detect planets by the discredited astrometric technique. Michael Shao was such a man. Shao had been thinking about using an interferometer in space to find planets ever since the 1970s. He had cut his teeth building and operating a sequence of progressively more capable ground-based interferometers on top of Mount Wilson, overlooking Pasadena, California. Shao came up with the idea of an Orbiting Stellar Interferometer (OSI) in the 1970s. OSI evolved through different designs and eventually became known as the Space Interferometry Mission. Shao's hope was that SIM would be able to perform an astrometric planet search with an accuracy that was at least 1000 times better than the best that had been done by ground-based telescopes. SIM would thus be able to look for planets as small as Earth.

NASA had formed the Planetary Systems Science Working Group in 1988 to guide its plans for discovering new planetary systems, and for the next 7 years the working group vigorously debated the merits of the ideas that had been paraded before it. These ideas ranged from the difficult to the seemingly impossible, including not only astrometric detection telescopes like Shao's, but also direct detection telescopes that might be able to snap the first picture of a planet around a nearby star. There was a third contender as well, dubbed FRESIP by its originator, William Borucki of NASA's Ames Research Center in the middle of Silicon Valley, California.

FRESIP, which stands for Frequency of Earth-Sized Inner Planets, was intended to find Earth-size planets on Earth-like orbits by watching a large number of stars and waiting for one of them to blink for a

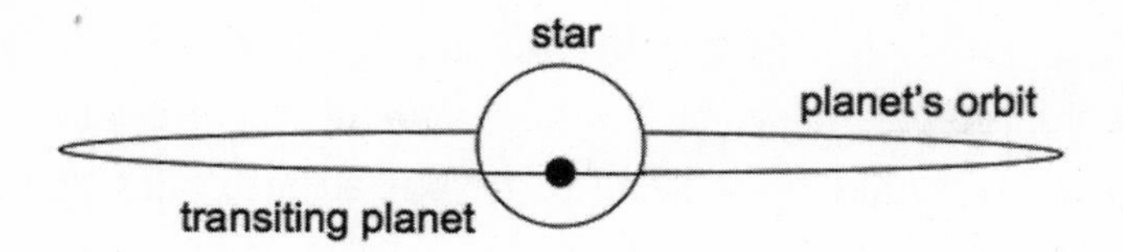

FIGURE 9. Planetary transits occur when a planet's orbit is aligned in such a way that the planet passes in front of and behind its star when viewed from Earth.

few hours as an Earth-like planet passed in front of, or transited, the face of the star, thereby dimming the star's light by a tiny amount, about 0.01%. This transit detection technique had been soundly thrashed in the Appendix of a 1990 National Academy of Sciences (NAS) report on the basis that FRESIP would have to look at tens of thousands of stars in order to have a chance of finding any planets at all, because of the small chances that the orbital planes would be aligned at precisely the right angles to yield transits—and there just weren't that many nearby stars.

Borucki was unperturbed by this published criticism, because he did not intend to use FRESIP merely to study nearby stars. He could point FRESIP at a rich field of stars somewhere in our Milky Way Galaxy and find the planets transiting in front of *those* stars. But others in the working group, including Robert Brown, lead author of the NAS Appendix, continued to resist the idea of a transit detection mission. Brown favored finding planets with Hubble, once it was equipped with a new camera specially designed to enable it to see planets by blocking out the light from the star. This would be done by an optical trick called coronagraphy, which is effectively the same as holding up your hand to blot out the Sun while searching the skies for flying birds. A coronagraph was basically a hand that was optimally shaped to block out the star's light, allowing the planets in orbit around it to be seen. Others in the working group argued for designing a coronagraphic telescope "from the ground up" to detect extrasolar planets, rather than trying to retrofit a space telescope like Hubble that had not been designed with planet detection in mind.

The working group spent years meeting, discussing, and debating the various ideas that were presented, and writing reports such as the TOPS plan, where the ever-present NASA acronym TOPS stood for Towards Other Planetary Systems. In terms of actually getting NASA to spend some serious money on planet detection, though, the working group accomplished little beyond getting NASA to pay for the countless boxes of donuts and gallons of coffee consumed at our meetings.

Finally, early in 1995, NASA headquarters asked Charles Elachi of Caltech's Jet Propulsion Laboratory (JPL) in Pasadena, California, to put together a plan—a road map—for systematically progressing from ground-based searches for Jupiters to space telescopes that might be able to find Earths. The lab had been given programmatic responsibility for extrasolar planets by NASA headquarters a few years earlier, and it wanted to get the planet detection program moving, not only for the payoff of finding new worlds, but also for the federal funds that were necessary to keep the laboratory running. Within just a few months, Elachi's team was ready to report its findings to the working group for review and approval. Elachi needed a new acronym for his road map, and he came up with ExNPS, standing for Exploration of Neighboring Planetary Systems. Elachi included in the plan the ongoing ground-based planet searches, such as Marcy and Butler's Doppler search at the Lick Observatory, and an astrometric search run by Gatewood at the Allegheny Observatory, because these programs were already under way, cost NASA comparatively little money, and hence would not interfere with Elachi's plans for a major effort. The ExNPS plan called for SIM to be built first, to prove the concept of interferometry on a boom in space, and to search the closest stars for Earths. But the *real* star of the ExNPS show was the plan to build an even larger space inteferometer than SIM, one 10 times longer, a boom perhaps 333 feet (about 100 meters) long with multiple mirrors strung out along its length. It would be called the Planet Finder.

The Planet Finder would operate at longer wavelengths than SIM, at so-called infrared wavelengths of light, similar to the light emitted by heat lamps that keep the french fries warm in a fast-food joint. Planet Finder would be an interferometer like SIM, but large enough to image extrasolar planets directly. The reason why Planet Finder would operate with infrared light, instead of visible light like SIM, was that stars give off much less light in the infrared than in the visible wavelengths, where their emission peaks. At the same time, planets such as Earth and Jupiter not only reflect some of the visible light they receive from the Sun but also emit their own infrared light. These two effects mean that although planets are about a billion times fainter than their stars at visible wavelengths, in the infrared they are only about a million times fainter.

However, because the resolving power depends on both the diameter of the telescope and the wavelength of light, an infrared telescope that operates at wavelengths that are 10 times as long as a visible-light telescope must be 10 times larger in order to have the same resolving power. Planet Finder would have to be 10 times larger than SIM. The payoff was that Planet Finder would be able to take pictures of the planets it found right away, obviating the need to wait years for the planets to show themselves through the astrometric wobbles of their stars. Given its huge size, whatever SIM cost, the Planet Finder could be expected to cost a lot more, in spite of their both being interferometers. That would mean steady work at JPL for a decade or more.

NASA Administrator Dan Goldin appeared on the second morning of the working group's final meeting at NASA headquarters on July 18, 1995, and spoke enthusiastically about the prospects for traveling the road laid out in Elachi's ExNPS road map. Compared to the TOPS report, which did not seem to have a clear destination in mind, the ExNPS road map led right to the ultimate goal of all of these efforts: new Earths. SIM would find them, and then the Planet Finder would take their pictures. What wasn't to like? The ExNPS road map envisioned that SIM would get under way immediately at the Jet Propulsion Lab in preparation for a launch in 2003. This meant that by 2004 or so, SIM would have determined the year-long orbits of the first Earth-like planets among the nearby stars.

Borucki's FRESIP did not make Elachi's cut and was not a part of the ExNPS road map. The road mappers were worried that random fluctuations in the light given off by stars would swamp the tiny dimming caused by transiting Earths, so they did not recommend developing a space mission like FRESIP. A few months later, NASA cut off the funding for FRESIP, as well as for the other competing mission concepts that did not make the final cut for the road map. Shao's SIM mission was already headed down the ExNPS highway, with FRESIP disappearing in the rearview mirror.

August 7, 1996—An electrifying claim hit the news media around the world: scientific evidence had been discovered for life on Mars. Dan Goldin had personally briefed President Bill Clinton and Vice President Al Gore about the discovery a week earlier in order to give the White House a chance to prepare for this momentous event. It was as if an alien spacecraft had landed on the Ellipse next to the White House and strange creatures had emerged asking to be taken to meet President Clinton.

David McKay and Everett Gibson of NASA's Johnson Space Center in Houston, Texas, led the team that analyzed a 5-pound piece of Martian rock. The rock had been blasted off the surface of Mars by a meteorite impact, putting the Mars rock on a trajectory that led it eventually to fall to Earth as a meteorite on the Antarctic ice sheet. A 1984 expedition to find Antarctic meteorites found the Martian rock near the Allan Hills, earning it the name ALH84001. The meteorite was carefully preserved and sent back to Johnson for curation, along with the lunar samples returned by the Apollo astronauts. Upon examination, minute traces of gas that duplicate what we know about the unique composition of the Martian atmosphere made it apparent that ALH84001 was a Martian meteorite, one of only about a dozen that have been found so far.

McKay and Everett found a number of hints that Mars had once been alive with miniscule nano-bacteria: bacteria-shaped tubes, organic compounds, magnetite crystals like those found in bacteria, and carbonate spheres produced by bacteria. Any one of the four hints would have been debatable, because they could have been formed abiotically by normal geological processes, but taken together they seemed to present a strong argument for Martian bacteria. The bacteria were clearly not up to the task of digging the putative Martian canals, but they would have been happy to live in one.

The discoveries of Mayor, Queloz, Butler, and Marcy had proved that other planetary systems exist in the universe. They had not found Earths, but the cool Jupiter in the 47 Ursa Majoris system looked like the type of Solar System analogue we were hoping to find, with room for inner habitable planets. If even chilly, frozen

Mars now sported evidence of life, and in particular if it could be proved that Mars life had arisen independently of Earth life, then there would be no question that life was prevalent in the universe. Plenty of planets seemed to be out there, and if life had arisen on at least two planets in our Solar System, there would be no reason why life could not be flourishing all over the Galaxy. The combination of the extrasolar planet discoveries and the claims for Martian life made for an intoxicating brew.

September 19, 1996—The Clinton White House released a new National Space Policy to guide NASA's future efforts in the context of the miraculous discovery of life on Mars the previous month. The plan called for NASA to send robotic spacecraft to Mars by 2000 to look for further evidence of life, and to bring samples back to Earth for more detailed analyses than could be done by a robotic Mars lander. The plan also charged NASA with discovering and characterizing planets in orbit around other stars. Suddenly the search for life in the universe had been pole-vaulted to the top of NASA's priorities by the Clinton White House. Former President George H. W. Bush's plan to return humans to the Moon and then send them on to explore Mars, announced with great fanfare in 1989, did not appear at all in the new plan for NASA.

October 17, 1996—The Space Telescope Science Institute held a workshop in Baltimore on extrasolar planets and how they could be studied with the next generation of space telescopes. Given that the jobs of the hundreds of astronomers at the Institute depended on the existence of Hubble, and that the telescope was scheduled to cease operations in 2005, the Institute astronomers were already looking around for a replacement space telescope to provide their daily bread. Ever since its founding in 1981, the Institute had embodied the "Full Employment Act" for astronomy, not only for those willing to work there but also for astronomers around the country, whose grants of valuable Hubble observing time were allocated in "orbits" (the usable portion of the roughly 96 minutes that it takes Hubble to circle Earth in an orbit low enough to be reached and serviced by the Space Shuttle). These grants of Hubble orbits were accompanied by sufficient research funds to make one question whether many astronomers were applying more for the Hubble orbits than for the Hubble dollars.

Bill Borucki had not given up on FRESIP. As is customary at NASA, when a mission or program is canceled or meets some similar fatal difficulty, one simply chooses a new name or acronym and gets on with business. Borucki gave a talk at the Baltimore meeting about the Kepler Mission, the new name for FRESIP. Rather than choose a new acronym, Borucki went with the rage at the time of naming missions after famous scientists, as had been the case when NASA's Space Telescope was named after the Carnegie Institution astronomer, Edwin P. Hubble, who had discovered the fact that the universe is expanding.

Johannes Kepler (1571–1630) was a German astronomer who built on the work of previous astronomers, such as the Polish astronomer Nicolas Copernicus, who proposed in 1530 that the planets orbited the Sun, not the other way around, and the Danish astronomer Tycho Brahe, who could not quite take the leap to believe in Copernicus's heliocentric, or Sun-centered, universe but nevertheless made sufficiently precise measurements of the locations of the planets in the sky to allow his successor, Kepler, to take the next logical steps. Kepler was able to solve many of the riddles of planetary motions, starting with his assertion in 1609 that the orbit of Mars

FIGURE 10. Johannes Kepler [1571–1630], the German astronomer who first understood the elliptical nature and basic laws of planetary orbits. [Courtesy of NASA's Kepler Mission and the Sternwarte Kremsmunster, Upper-Austria.]

was elliptical, with the Sun lying at one of the foci of the ellipse. (In reality, the center of mass of the planet-star system lies at one focus of the ellipse, but Brahe's observations were not precise enough for Kepler to make such a fine distinction.) Kepler's Third Law of planetary motion showed how the orbital periods of planets are related to their distances from the Sun, a fact that is used routinely to determine the sizes of the orbits of extrasolar planets where only the orbital period is determined directly. Kepler also founded the field of astrometry, which Peter van de Kamp would use over three centuries later to search for a planet around Barnard's star. And Kepler accomplished all of this during an era when he had to spend time defending his mother against charges of witchcraft.

Borucki estimated at the Baltimore meeting that if the Solar System was typical of planetary systems, the Kepler Mission would discover not just one or two but hundreds of Earth-like planets. If Borucki was correct, the Kepler Mission would tell us just how common Earth-like planets were, unless it found none at all, in which case all we would have would be an upper limit on their frequency. But even an upper limit would be important to know in order to estimate the overall probabilities for life in the universe.

February 6, 1997—President Clinton launched the Origins Initiative, intended to study the origin of life in the context of the formation of planets, stars, and galaxies. This initiative included plans for planet-finding space telescopes and a new emphasis for NASA on astrobiology, an interdisciplinary science that would seek to understand how life begins and evolves. A Mars Sample Return Mission was planned, giving Earth-bound laboratories a chance to subject Martian rocks to the same scrutiny as that to which the Apollo lunar rocks had been subjected. This time the mission would be accomplished much more inexpensively and safely by a robot, not by NASA astronauts. With White House support, the scientific wish list began to grow apace, with other agencies besides NASA, such as the National Science Foundation, wanting a piece of the new pie. The Administration's Office of Management and Budget, following the White House directive, promised to give NASA an extra $1 billion over the next 5 years to spend on the Origins Initiative. The big-ticket items were the Mars Sample Return Mission, SIM, and Elachi's Planet Finder.

April 9, 1997—Deane Peterson of the State University of New York at Stony Brook and chair of the SIM working group, e-mailed me to say that SIM was now a line item in the president's budget for the Origins Initiative. Thereafter the project would be more or less guaranteed a stable source of funds and hence would not be so vulnerable to the day-to-day funding crises that can rob an infant project of its lifeblood. The SIM project was cost-capped at a total cost of $450 million, not including the cost of the launch vehicle. Serious work on SIM would start at the Jet Propulsion Lab in October 1997, and SIM's launch was scheduled for August 2004. With any luck, we would know whether there were Earths around the closest stars just a few years after SIM's launch, perhaps by 2006, less than 10 years away.

October 29, 1997—The SIM working group held a meeting at JPL, where we learned that SIM was going to be the largest project at JPL for the next decade. The total cost to develop, build, and launch SIM had now risen to about $600 million, up from the number in the presidents' budget for 1997–1998 of $450 million. SIM's launch date had also slipped from 2003 to 2005. The start of serious work on SIM at JPL had been delayed by nine months because there wasn't enough money at NASA to get SIM started and also to begin work on a competing space telescope, the Next Generation Space Telescope that was intended to replace the Hubble Space Telescope.

The jump in SIM's cost estimates, which delayed the launch date by 2 years, was ominous, but given the euphoria over the fact that SIM was now a line item in the president's budget, a slip of a few years did not seem too terrible a burden to bear. SIM was going to find nearby Earths, and that was all that mattered.

CHAPTER 3

Kiss My Lips, Tootsie

It is a capital mistake to theorize before one has data. Insensibly one begins to twist facts to suit theories, instead of theories to suit facts.

—[SIR] ARTHUR CONAN DOYLE (1859–1930)

It is also a good rule not to put overmuch confidence in the observational results that are put forward until they are confirmed by theory.

—[SIR] ARTHUR STANLEY EDDINGTON (1882–1944)

January 7, 1998—NASA administrator Dan Goldin addressed thousands of astronomers at a meeting of the American Astronomical Society, this time in the International Ballroom of the Capital Hilton in Washington, D.C. Elachi's Planet Finder had been renamed the Terrestrial Planet Finder, or TPF, just to make it clear exactly what the goal was: new Earths. Goldin was bursting with enthusiasm for what the White House's endorsement of the Origins Initiative could mean for NASA's future. TPF would be launched in 2008, Goldin said. NASA was currently under an agency-wide hiring freeze, and its effort to recruit biologists who were willing to join NASA and become astrobiologists was the one exception. Goldin said that if TPF found nearby Earths, then NASA would plan to build an even more ambitious space telescope capable of imaging the surfaces of these newfound worlds, with enough resolving power to search for signs of ice caps, clouds, continents, oceans, maybe even interstate highways and their ubiquitous signs for "South of the Border."

Goldin envisioned that such a Planet Imager telescope would be a gigantic version of TPF itself; another space infrared interferometer, but its mirrors spaced 3800 miles (about 6000 km) apart, rather than 333 feet. The Planet Imager would fly in 2020, just over two decades away. Goldin concluded his address by suggesting that NASA would want to send robotic spacecraft to the closest habitable worlds by the end of the twenty-first century in order to take really close-up photographs of the planets and beam the images back to us at the speed of light. Goldin's breathtaking proposals were enough to keep NASA's engineers gainfully employed for the next century, giving them something much nobler to do than run a shuttle service for a chosen few from the east coast of Florida to low Earth orbit and back.

June 8, 1998—Maryland Senator Barbara Mikulski announced that Baltimore's Space Telescope Science Institute had been chosen to operate the Next Generation Space Telescope, an infrared space telescope with a diameter three times that of Hubble. Hubble itself had been given a 5- year reprieve and was now scheduled to be de-orbited by a Space Shuttle mission in 2010. Letting Hubble fall on its own to Earth risked the public relations disaster of its massive 94-inch primary mirror crashing down in an inhabited area, as pieces of NASA's Skylab space laboratory had done in Western Australia in 1979. The Australian chickens that were startled by the Skylab impacts refused to lay eggs for a week. Hubble's primary mirror would make a more impressive crater than the chunks of Skylab did, potentially inflicting even more severe losses on local egg producers.

The Next Generation Space Telescope was planned for launch in 2007, providing a 3-year overlap during which Hubble and the Next Generation Space Telescope could observe the same astronomical objects, allowing the Next Generation Space Telescope's infrared sensors to detect features that Hubble could not, given Hubble's primary emphasis on visible light. The Space Telescope Science Institute's astronomers cheered Senator Mikulski's announcement. Mikulski also announced that the Next Generation Space Telescope would be built at NASA's Goddard Space Flight Center in Greenbelt, Maryland, even though NASA headquarters said that where the Next Generation Space Telescope was to be built had not yet been decided. As the frequent chair or ranking minority member of the U.S. Senate Appropriations Subcommittee with jurisdiction over NASA's budget, Senator Mikulski was used to having her voice heard loud and clear. NASA headquarters later announced that NASA Goddard had indeed won the competition to build the Next Generation Space Telescope.

The appointment of Hubble's successor was the direct result of a study released in 1996 by an independent committee of astronomers chaired by Carnegie Institution astronomer Alan Dressler. Dressler's report concluded that the two main goals of such a telescope should be to study newly formed galaxies and to detect Earth-like planets around other stars. However, when the Space Telescope Science Institute and NASA Goddard astronomers had a chance to write their own report in 1997, the newly named Next Generation Space Telescope was focused only on finding primeval galaxies and had misplaced the planet-finding goal previously established for it by Dressler's group. The segmented design chosen for the Next Generation Space

Telescope, a pattern of hexagonal mirrors like that of the Keck telescopes, meant that the Next Generation Space Telescope could never do the job of detecting Earths: there would be too much stray starlight coming from the edges of the hexagonal mirrors for the planet's much fainter light to be seen.

It was clear that many astronomers still did not think that extrasolar planets were objects worthy of serious study, in spite of the excitement initiated by the discovery of 51 Pegasi's planet.

June 22, 1998—Geoff Marcy announced the first detection of a planet around a red dwarf star, Gliese 876, at an International Astronomical Union (IAU) meeting at the University of Victoria, in British Columbia, Canada. Marcy and Butler, now at University of California–Berkeley and the Anglo-Australian Observatory in Australia, respectively, had used the Doppler method to show that Gliese 876 was circled by a presumed gas giant planet with a minimum mass of 2.1 Jupiter masses, orbiting with a period of 61 days at a distance of 0.21 AU on an eccentric orbit. Gliese 876 was located only 15 light-years away from Earth, making its planet the closest one to Earth found to date—by a factor of 3.

Two hours later, Marcy was handed an e-mail saying that Xavier Delfosse of Michel Mayor's Geneva Observatory team had independently found the same planet around Gliese 876. The next day, Mayor sent out an e-mail through French astronomer Jean Schneider's exoplanet e-mail exploder list, stating that Delfosse had e-mailed news of his probable discovery to Mayor on June 18. Delfosse was continuing to take data on Gliese 876 during an observing run at the European Southern Observatory at La Silla in Chile. By June 22, the day of Marcy's announcement, Delfosse was convinced that Gliese 876 had a planet and e-mailed Mayor again with the good news. But

Marcy had beaten the Swiss–French team to the podium at the Victoria meeting. Butler and Marcy also beat the Geneva Observatory team in the race to submit their Gliese 876 b discovery for publication in a refereed journal. The IAU's Working Group on Extrasolar Planets, which I was to chair during 6 of its 7 years of existence, decided to award the discovery rights for extrasolar planets on the basis of the date on which the paper was received for consideration by the journal. The paper that Marcy, Butler, and their colleagues submitted to the *Astrophysical Journal* was received on July 7, 1998, whereas the paper sent to the European journal *Astronomy & Astrophysics* by Delfosse, Mayor, and their colleagues was not received until August 4, 1998. One month had become an eternity in the ongoing battle between the American–Australian and Swiss–French teams for planet discovery bragging rights. The box score now read 8 to 4, with the American–Australian team having a strong lead in spite of the fast start by the Swiss–French team with 51 Pegasi b a few years earlier.

Gliese 876 is a red dwarf star of the M spectral type. Astronomers classify stars by the emission and absorption lines seen in their spectra, and they use these lines, in concert with theoretical models, to determine the masses of the stars. Higher-mass stars have hotter atmospheres than lower-mass stars and so have different spectral lines. For historical reasons, astronomers use letters of the alphabet to designate these spectral types, the ordering from high-mass to low-mass stars being O, B, A, F, G, K, M. To remember that peculiar sequence of letters, astronomers use as a mnemonic device the phrase "Oh, Be A Fine Girl, Kiss Me." The recent discovery of two more classes of even lower-mass stars, with spectral types L and T, has led the mnemonic to be changed to "Oh, Be A Fine Girl, Kiss My Lips, Tootsie."

The fact that M dwarf stars have eccentric planets was a revelation for the overall frequency of planetary systems. M dwarf stars like

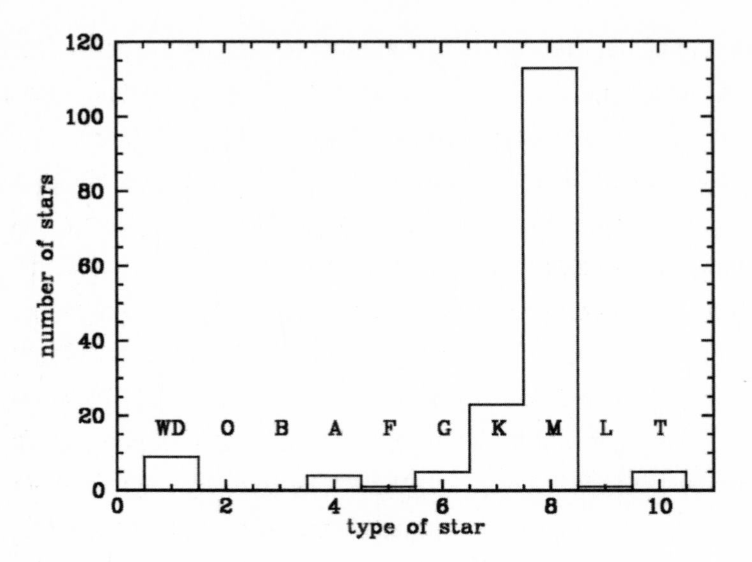

Figure 11. Census of the number of stars of different types [O, B, A, F, G, K, M, L, T] and white dwarfs [WD] within 25 light-years of Earth visible from the Northern Hemisphere. M dwarfs dominate the stellar population. [Adapted from data provided by J. Davy Kirkpatrick (Caltech).]

Gliese 876 and Barnard's star are by far the most numerous stars in the Galaxy; 113 of the 161 stars known to lie within 26 light-years of the Sun are M dwarf stars. The closest stars are mostly M dwarfs, with masses between one-half and one-tenth that of the Sun. Gliese 876 has one-third the Sun's mass. Mass-challenged stars like Gliese 876 would be attractive targets for searching for life on other nearby worlds. Now we knew that they had gas giants, but what about Earths?

February 11, 1999—The SIM Science Working Group held a meeting at the University of Arizona's Steward Observatory, in Tucson,

Arizona. SIM was still planned for launch in 2005, and the debate centered on whether SIM should be able to use its formidable diameter of 33 feet to take images or just to measure astrometric wobbles. To be of most use to the general community of astronomers, whose support was highly desirable, SIM should be able to take pictures of distant stars and galaxies four times sharper than those provided by the Hubble telescope. However, because SIM was an interferometer, not, like Hubble, a so-called filled-aperture telescope with a single primary mirror, SIM would have to be able to move its mirrors along its 33-foot boom, or else have mirrors placed along the boom at irregular intervals, in order to provide an image that would not look as if it had been taken by a camera with gobs of Vaseline smeared all over its lens. In order to find planets astrometrically, all SIM needed was the two mirrors on either end of the boom—the other mirrors needed for imaging added cost and complexity, and, worst of all, mass. Perhaps the greatest danger in building robotic spacecraft is designing a beast too heavy to be lifted into orbit. Cost overruns could always be paid for by robbing other projects in the agency, but a telescope too heavy to be lifted into space would be the ultimate show-stopper.

The SIM Science Working Group debate about whether to take images or just measure wobbles went on and on, and eventually it was put to a straw vote. The balloting yielded 8 votes in favor of imaging and 8 voted opposed: a tie. This meant that the Working Group's advice to JPL and NASA headquarters was basically "Do whatever you want." The Working Group members joked among themselves by drawing parallels between their vote and the ongoing debate in the U.S. Senate about whether to impeach President Bill Clinton over the Monica Lewinsky affair, which had been under way at the White House around the time of the ALH84001 Martian rock sensation. White House political advisor Dick Morris had told the secret about the possible discovery of life on Mars to his mistress, which she had mistakenly recorded in her private diary as life on Pluto.

February 20, 1999—The SIM Project at JPL decided to retain the option of astronomical imaging, if for no other reason than to provide a convenient "descope" option when SIM inevitably ran into trouble at some time in the future. NASA headquarters concurred. By planning for SIM to have this capability, the project could get the development funds needed for imaging right away but later spend them elsewhere if necessary. Space projects often hedge their bets against future technological hitches and cost overruns by setting aside a fixed unassigned percentage of the expected total costs as "reserves" and by proposing enough capability up front for some of that capability to be bargained away at a later date, if necessary.

September 16, 1999—*Nature* reported that a planned space telescope named CoRoT was in danger of being dropped by the French national space agency, the Centre National d'Etudes Spatiales, or CNES, because of budget cuts imposed by the French science ministry. CoRoT, which stands for Convection, Rotation, and Planetary Transits, was conceived around 1994 by French astronomers who wished to learn more about the interior structures of stars by monitoring their vibrations, in somewhat the same way that Earth's core/mantle structure can be discerned by studying the planet's vibrations—that is, the seismic waves generated by earthquakes. CoRoT was planned as a stellar seismograph, able to probe deep into stellar interiors. Because CoRoT would be staring at fields of stars for as long as 150 days, however, it might also be able to detect planetary transits, at least for planets with orbital periods much less than 150 days. Multiple transits could be observed and the transiting planet's orbital period reliably determined. Frederic Bonneau, CoRoT's head at CNES, noted that CoRoT would also be able to determine the physical size of transiting planets by measuring how much of the target star's light was blocked by the area of the planet during a transit: a planet 10 times smaller than its star would lead to a 1% diminution in the star's light, because the areas of the planet's and star's surface projections on the plane of the sky depend on the squares of their respective diameters. CoRoT aimed to produce as accurate a picture of transiting planets as one might imagine could be created by the French realist painter Jean-Baptiste-Camille Corot (1796–1875).

CoRoT was planned for launch in 2002 with a total cost of only $55 million, including a Russian launch vehicle—a bargain compared to the expected costs of a typical NASA space mission. However, even this low total cost put CoRoT in jeopardy, given the desire of Claude Allègre, the French science minister, to cut major projects. Allègre is a well-known and respected cosmochemist who had helped fix the age of the Solar System at 4.566 billion years, give or take a few million years. However, Michel Mayor had confided to me at a 1998 exoplanets meeting in Lisbon, Portugal, that Allègre had wanted to close the Haute Provence Observatory, where Mayor had discovered 51 Pegasi b, and to cut the budget for the European Southern Observatory by 25%. Now Allègre wanted to finish the job on French astronomy by sending CoRoT to the guillotine. Apparently Allègre was content with planet Earth and felt no need for France to continue to discover new worlds.

October 22, 1999—The estimated cost of SIM rose to the range of $650–$700 million, well above the NASA headquarters target of $500 million. The SIM Project was told that it must cut costs or risk being canceled so that its funds could be used to help build the Next Generation Space Telescope, which now had a total cost estimate of $2 billion and a launch date of 2008. Clearly the Next Generation Space Telescope was the favored child, in spite of SIM's ability to perform brilliantly on a number of tough astronomical problems, such as helping to refine the cosmic distance scale, and in spite of SIM being NASA's first planet-hunting space telescope.

November 11, 1999—Paul Butler was excited, very excited. His team had made a major discovery, arguably the most important discovery in the field of extrasolar planets since 51 Pegasi's planet appeared in 1995. Butler, normally a phlegmatic person, decided to let me in on his team's bombshell. Butler had become a colleague a few months earlier, having left Australia for a staff appointment at the Department of Terrestrial Magnetism—and for the steady access to the Carnegie telescopes at the Las Campanas Observatory in Chile that this appointment entailed. A true astronomer lives for telescope time, and Butler is a true astronomer, typically spending half of his time at observatories in Australia, Chile, and Hawaii or in transit somewhere in between one of those three locations and Washington, D.C.

The Doppler planet search teams had found nine hot Jupiters so far, yet they had not found a single one that was a transiting planet, which was unsettling. The closer a planet is to its star, the more likely it is that a randomly inclined orbit will result in the planet periodically blocking some of the star's light when it passes in front of the star as seen from Earth. For hot Jupiters, there is a 10% chance that a transit will occur. Because Jupiter has one-tenth the diameter of the Sun, a hot Jupiter should produce a transit where the star's light dims by about 1%, a small dip but one that is readily measured by ground-based telescopes. The puzzle was, with nine hot Jupiters known, why had none as yet been found to transit?

Could it be that the hot Jupiters were not really gas giants similar to Jupiter? Perhaps they were more like the pulsar planets, made not of hydrogen and helium gas but of something else, giving them such a high density and small diameter that the dimming produced by their transits could not be seen from Earth. Worse yet, maybe their orbits were inclined in such a way that transits did not occur. But if that were true, it would mean that their true masses could be much higher than the minimum masses derived from the Doppler detections. In that case, they might not be planets at all but very low-mass stars called brown dwarfs. Such L and T dwarfs are interesting objects in their own right, but they are certainly not planets.

The bombshell was that Butler's team had found the tenth hot Jupiter, and this tenth one transited. The estimated 10% chance for transiting hot Jupiters had held true, but we had had to roll the dice 10 times to finally get the expected payoff.

Butler's team had added a star called HD 209458 to its observing list just a few months earlier, in May 1999. HD 209458 was the 209,458th entry on a list of over 225,000 stars compiled in the late 1800s by American astronomer Henry Draper. As of May, it was one of a list of about 500 stars being carefully monitored by Butler's team with the HIRES (High Resolution Echelle Spectrometer) spectrograph on the Keck I telescope in Hawaii. HIRES had been designed by team member Steven Vogt, who was Marcy's Ph.D. thesis advisor at UC Santa Cruz. Vogt's superior skill in building HIRES allowed their California-Carnegie team to routinely measure Doppler shifts with errors of only 7 miles per hour (3 meters/second) at the Keck Observatory on Mauna Kea.

A few days earlier, on November 5, Butler's analysis of the data for HD 209458 revealed that it had a hot Jupiter with an orbital period

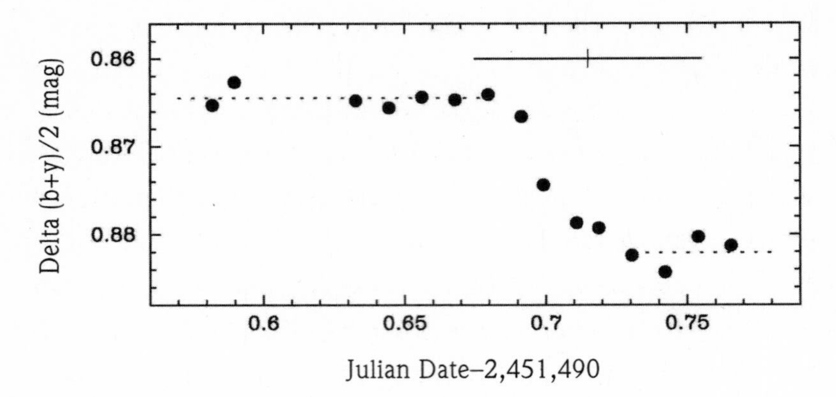

FIGURE 12. First detection of a transiting planet, orbiting HD 209458, using the T8 automated telescope of Fairborn Observatory, Arizona. Only the first portion of the transit was observed, showing the star's light dimming by about 1.6% as the planet passes in front of the star. [Reprinted, by permission, from G. Henry et al., 2000, *ApJ*, volume 529, page L42. Copyright 2000 by the American Astronomical Society.]

of 3.5 days. Once every 3.5 days, there was a chance that a transit could be seen as the hot Jupiter passed through the point on its orbit closest to Earth, the point known as "inferior conjunction." Butler's data analysis showed that the next such chance would occur on the night of November 7, so team member Gregory Henry of Tennessee State University was asked to monitor HD 209458 at the appropriate time that night, using the automated 32-inch (0.8-meter) T8 telescope at the Fairborn Observatory in Arizona. He did so, and he saw clear evidence for the beginning of the transit, as HD 209458 dimmed by 1.6% at the predicted time. However, about 2 hours after the transit began, Earth's rotation had carried HD 209458 so low in the sky that Henry's ability to measure the star's brightness accurately was degraded by the increasing amount of atmosphere the star's light had to pass through before reaching the T8 telescope. The next transit, 3.5 days later, occurred during the day and so could not be observed, and on the night of the third transit, 7 days after the first, the skies were cloudy in Arizona. Even a small amount of clouds makes precise measurements of stellar brightness impossible, because the varying thickness of high cirrus clouds passing in front makes the star seem to disappear and then reappear. The team would have to rely on the data from the first half of the November 7 transit.

On November 11 the California-Carnegie team decided to alert the rest of the world to the transiting planet in the hope that someone else would be able to observe the entire transit. Following standard astronomical procedure, Henry asked Brian Marsden of the Smithsonian Astrophysical Observatory to send out an IAU Circular announcing the discovery of the transit and predicting further transits on November 15, 18, and 22. By the next day, the IAU Circular web page still had not shown the discovery, so Marcy called a few astronomers to give them the news directly, including Timothy Brown of the High Altitude Observatory in Boulder, Colorado. Brown was not in his office, so Marcy left a message. Brown returned the call later in the day, and it was then that Marcy learned Brown's group also had evidence for a transit.

Michel Mayor's team had been observing a star called SAO 107623 at the Haute Provence Observatory, while David Latham of the Harvard-Smithsonian Center for Astrophysics in Cambridge, Massachusetts, had been observing the star with HIRES on Keck I. Mayor and Latham were old friends who had collaborated in 1988 in discovering a possible planet around the star HD 114762 with a minimum mass of 11 Jupiter masses, which had made them think it was more likely to be a brown dwarf than a gas giant. Latham and Mayor pooled their data, took more data with the 48-inch (1.2-meter) Swiss telescope at the European Southern Observatory's facility at La Silla in Chile, and then realized they definitely had a new hot Jupiter. It was time to check it for a transit.

Looking for a transit fell to Tim Brown and Harvard graduate student David Charbonneau. In July 1999 Brown had installed a homemade 4-inch (10-cm) telescope in a shed near the parking lot of his office building in Boulder, Colorado. Prompted by Latham, Charbonneau monitored the brightness of SAO 107623 on 10 nights between August 28 and September 15. Charbonneau began checking the data for evidence of a transit on October 27, and by November 10 he and Latham had decided they had found the first transit.

On November 12, Brown stopped by his office and found the telephone message Marcy had left earlier that day. He returned Marcy's call, suspecting that both teams had made the same discovery. Brown and Marcy soon figured out that Marcy's star HD 209458 was the same as Brown's star SAO 107623—bingo! Each group's data provided independent confirmation of the reality of the transit of HD 209458 b. There was no way that they could both be wrong.

After Marcy hung up, he decided it was time to change the plan for the press release his team had prepared. Anne Kinney, head of the Origins Program at NASA headquarters, had wanted to trumpet the discovery with a formal press conference on November 19, given that the California-Carnegie team's discovery had been made possible in part by telescope time awarded through NASA's one-sixth share of the Keck Observatory. Now that there was a danger that the Geneva/Harvard-Smithsonian/Boulder team would scoop them, Marcy decided to jump in first. For all he knew, there might be other teams out there similarly poised to claim the discovery. After losing out to Mayor and Queloz on the discovery of the first extrasolar Jupiter, Marcy and Butler were not about to let this prize slip through their fingers. The press release was sent out the night of November 12 to Stephen Maran, the American Astronomical Society Press Officer, for distribution to the world's media. That night Butler and Marcy began feverishly writing the paper that would cement their claim for discovery of the first transiting planet.

By November 14, the California-Carnegie team's discovery was being ballyhooed on television shows and on newspaper front pages across the country, with no mention of the essentially simultaneous discovery by the Geneva/Harvard-Smithsonian/Boulder team.

November 18, 1999—The *Astrophysical Journal* office received the paper by Henry, Marcy, Butler, and Vogt describing the combined Doppler and transit detection of HD 209458 b. The following day, the *Astrophysical Journal* received the paper by Charbonneau, Brown, Latham, and Mayor with their transit data. By the rules of the IAU Working Group on Extrasolar Planets, the discovery rights went to the Henry paper, although the *Journal* wisely published the papers back to back, Henry's first, in the January 20, 2000, issue of its Letters section.

The discovery of this first transiting planet was a milestone for a number of reasons. Notably, it was the first time that a planet found via Doppler wobbles had also been found by a completely different technique, transit photometry. Doppler had thus been confirmed to work as advertised. Second, because the planet's orbit had to be seen nearly edge-on in order to produce a transit, the minimum mass found by the Doppler detection had to be close to the true mass of the planet. In the case of HD 209458 b, the planet's mass was nailed down at 0.69 Jupiter mass, less than Jupiter's mass but above Saturn's mass of 0.30 Jupiter mass. Third, this first true mass strongly indicated that HD 209458 b was indeed a gas giant, like Jupiter and Saturn, and presumably so were most of the other planets found by the Doppler surveys. Fourth, and perhaps most important for making the

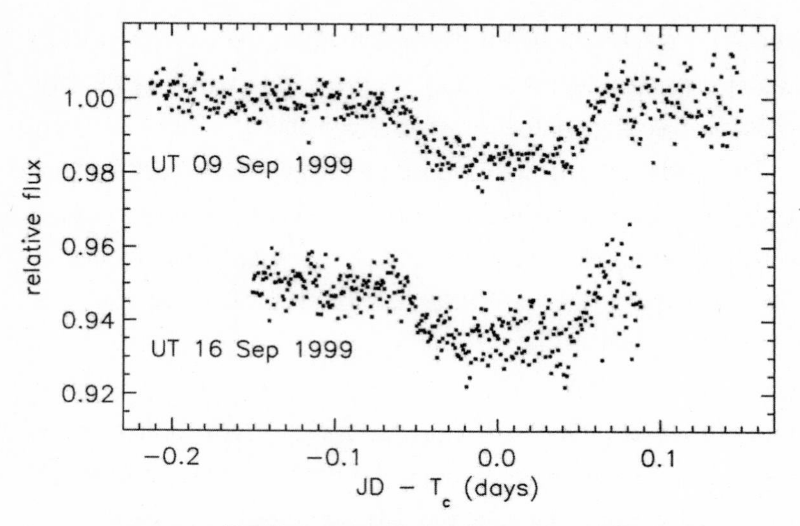

FIGURE 13. First detection of a transiting planet, orbiting HD 209458, using a 4-inch telescope in a parking lot in Boulder, Colorado. The entire transit event was observed on two different nights. [Reprinted, by permission, from D. Charbonneau et al., 2000, *ApJ*, volume 529, page L45. Copyright 2000 by the American Astronomical Society.]

case for giant planets, the transit depth of a few percent meant that the radius of the planet could be determined to be about 30% greater than that of Jupiter, implying an average density less than that of either Jupiter or Saturn. And with that low a density, HD 209458 b had to be a gas giant planet, probably "fluffed up" because it was a hot Jupiter.

November 22, 1999—Bill Borucki's Kepler team was able to confirm the transit of HD 209458 b using their Vulcan camera at the Lick Observatory, whose domes are visible on Mount Hamilton to the south of NASA Ames. In order to test out Kepler's transit photometry technique, Borucki and his NASA Ames colleagues had built a small camera able to monitor 6000 bright stars simultaneously, searching for the periodic dimming that signals a planetary transit. Vulcan had yet to find a new planet, but once HD 209458 b's transit was announced, Borucki knew where and when to point Vulcan, and Vulcan did the rest. Transit photometry worked, and therefore so should Kepler.

November 23, 1999—Dave Charbonneau's paper on HD 209458 b's transit was accepted for publication by the *Astrophysical Journal.* Tim Brown used the exoplanet e-mail exploder to announce the team's results, following the standard scientific etiquette of waiting for an important paper to be accepted (or even published, depending on the journal's policy about embargoes) before making the hot news public.

May 18, 2000—The new NAS Decadal Survey on astronomy and astrophysics was issued. Commissioned every 10 years by NASA and the National Science Foundation, these reports help U.S. astronomers decide among themselves what their priorities should be for the expensive ground- and space-based telescopes that determine the future of their science. SIM had previously been blessed by the 1991 Survey, and the 2000 Survey therefore assumed that SIM would fly and simply reaffirmed its support for SIM. The real question was what the 2000 Survey would say about the Terrestrial Planet Finder.

The 2000 Survey recommended TPF but gave it a relatively low priority: TPF was sixth on a list of seven major initiatives, with the Next Generation Space Telescope at the top. The Survey insisted that TPF should not only be able to find planets but should also be capable of taking images in the infrared that were 100 times better than any ever taken before. Evidently the demand for exquisite imaging was what motivated the general astronomical community to rate TPF even at number six on the list. Neverthless, Anne Kinney, who was responsible for SIM and TPF, optimistically said, "I'm delighted with this report—it's fantastic." And NASA headquarters apparently took the fact that TPF was in the 2000 Survey as sufficient support to proceed with the $1.7 billion mission. TPF remained a favorite of administrator Dan Goldin, and that fact could not be discounted in what NASA would do.

October 5, 2000—CoRoT was approved for launch in 2004 by CNES. Claude Allègre had been fired as the French science minister in March after a 3-year reign of terror for French scientists. CoRoT had survived, and it had gained the support of ESA (European Space Agency) as well, making it a more truly European Union space mission.

The transiting planet space race with Kepler was on, but only if Kepler could win the support of NASA headquarters. One way to do this was to get Kepler chosen as a Discovery-class mission, where teams of scientists propose specific missions of their own design. Bill Borucki had entered Kepler in the 1998 competition for a NASA Discovery Mission slot but had failed to win anything except some much-needed technology development funds. With a new Discovery competition under way in 2000, Borucki was ready to try again. CoRoT's approval might mean that NASA would feel there was now no need for Kepler, *or* it could stimulate a flow of competitive juices that would help push Kepler forward.

The 2000 Decadal Survey had demanded that before TPF could be "started," astronomers had to "confirm the expectation that terrestrial planets are common around solar-type stars," which is exactly what Kepler would do.

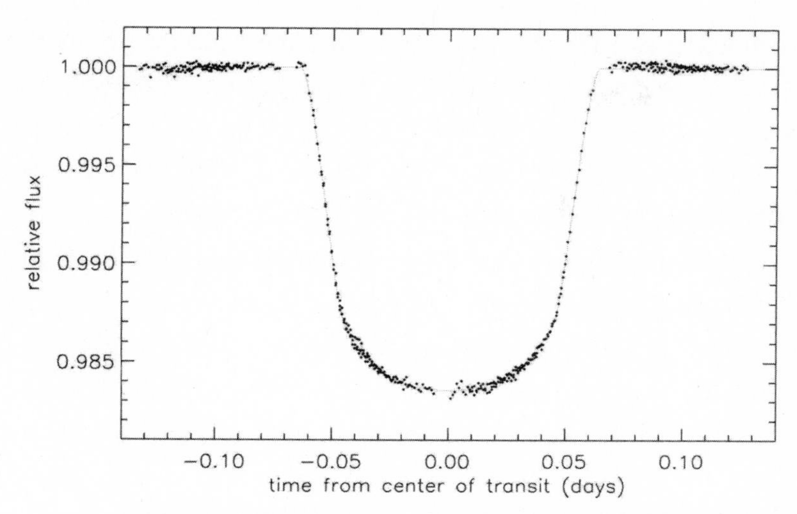

FIGURE 14. **Confirmation of the first transiting planet, HD 209458 b, by the Hubble Space Telescope. This image illustrates the great advantage of observing transits from above Earth's atmosphere, which results in there being much less noise in measuring the brightness of the star. [Reprinted, by permission, from T. Brown et al., 2001, *ApJ*, volume 552, page 702. Copyright 2001 by the American Astronomical Society.]**

October 25, 2001—David Charbonneau and Tim Brown submitted a paper to the *Astrophysical Journal* about using Hubble to detect the presence of sodium atoms in the atmosphere of HD 209458 b, by observing minute changes in the star's light at the frequencies absorbed by sodium atoms during the transit. As the planet passed in front of the star, some of the star's light had passed through the planet's atmosphere on its way to Hubble, and this passage had left a faint but discernible pattern on HD 209458's light. Charbonneau and Brown and their colleagues had accomplished the first characterization of the composition of an extrasolar planet, without any question of their priority for this major discovery. This feat proved that space telescopes could reveal the chemical composition of exoplanets, buttressing the case for TPF's goal of not only detecting nearby Earths but also telling us enough about their atmospheres so that we could guess whether they were habitable—or maybe even inhabited.

November 14, 2001—Following the 2000 presidential election debacle, where Al Gore won the popular vote but George W. Bush was effectively appointed president by the U.S. Supreme Court, Office of Management and Budget Deputy Director Sean O'Keefe was nominated by President Bush to succeed Dan Goldin as head of NASA. Bush felt that a change in NASA's leadership was necessary "to ensure fiscal responsibility." O'Keefe had been a bean counter for the Navy and the Department of Defense, with close ties to Vice President Dick Cheney, so he got the nod from Bush, or Cheney, or whoever decided these things in the Bush White House. On November 16, 2001, Dan Goldin stepped down as head of NASA.

December 11, 2001—The TPF Final Advocate Review meeting was held on Shelter Island in San Diego Bay, California. The Jet Propulsion Lab had selected four teams of engineers and scientists to start fresh and come up with new approaches for the design of a space telescope capable of finding living planets. The goal was to consider designs other than the two that had been studied to date. Elachi's Planet Finder, the 333-foot infrared interferometer on a stick, had been succeeded in JPL's plans by the idea of putting four 140-inch (3.5-meter) telescopes on four separate spacecraft, so that the four mirrors could be flown around to form an interferometer that had the resolving power of a telescope with a diameter anywhere from 248 feet (75 meters) to 0.6 mile (1 kilometer). Such a free-flying version of TPF was the same basic design as that being pursued by the Europeans, whose mission was named Darwin. TPF/Darwin was going to be so expensive that it was clear that both NASA and the European Space Agency should plan to collaborate and share the enormous costs, as they had done with Hubble. ESA's budget is roughly 25% of NASA's budget, so ESA was particularly motivated to collaborate on TPF/Darwin. NASA was willing to collaborate but hoped to take the lead, as it had done with Hubble.

At the San Diego meeting, Origins Czarina Anne Kinney said that "All roads lead to TPF." Noting the 2000 Decadal Survey language, Kinney also pointed out that the Kepler Mission, currently under review for a Discovery Mission slot, would be able to provide the estimate of the frequency of Earth-like planets that the Survey considered a prerequisite for "starting" TPF. Knowing this frequency would also be important for choosing the optimal design for TPF: if Earths were commonplace, then TPF need search only the closest stars to find some, whereas if Earths were relatively rare, TPF would have to be able to search much deeper in space, and so might require a different design. JPL planned to "start" TPF by 2007 with a launch planned for 2014. Cost estimates varied widely, but using the total mass as the most important parameter led to estimates in the range of $1–$2.5 billion.

NASA headquarters and JPL decided that the Decadal Survey's requirement for exquisite astrophysical imaging could be safely ignored. There was thus no need to insist on the free-flyer concept for TPF, which would be the only way to get that kind of resolving power.

The major outcome of the San Diego review was the realization that TPF could be built to work at visible wavelengths, not just at infrared wavelengths, as had been assumed previously. In spite of an Earth being roughly 10 billion times fainter than the star at visible wavelengths, compared to 10 million times fainter at infrared wavelengths, the clever astronomers on several of the teams decided that they could make TPF work with visible light. They would do this by building a space telescope with a single primary mirror that had a diameter of 320 inches (8 meters) and use a coronagraph to block out the star's light, allowing the planet to be seen. Building such a coronagraph would not be easy, but enough progress had been made in optical science to make it seem not impossible. Coupled with this revelation was the realization that visible light offered the same sort of atmospheric biomarkers that infrared light offered for taking the next step beyond detection: characterization of the molecules in the atmospheres of the detected planets (such as oxygen, water, and carbon dioxide) was necessary to determine whether the planets were likely to be habitable. If methane were found as well as oxygen, there would be a good chance that the planet was alive; the methane in Earth's atmosphere is largely a product of methanogenic bacteria and bovine belching and flatulence. Any detection of methane on an extrasolar Earth would therefore be of immense interest to microbiologists, as well as to the American Cattlemen's Association.

As a result, TPF-C was born at the San Diego meeting. The free-flying infrared interferometer version was still a serious option as well, and it would now be known as TPF-I. If Earths were commonplace, then either TPF-C or TPF-I could be successful, but if Earths were few and far between, TPF-I would be necessary. JPL planned to decide between the two architectures for TPF by 2006 and to continue to develop only a single design thereafter. The 2006 downselect would be based primarily on the technological readiness of TPF-C versus TPF-I, because it was certain that the frequency of Earths was not going to be established by 2006. Borucki's 2000 Discovery Mission proposal for Kepler planned a launch in late 2005, and it would take the next 5 years to derive a good estimate of the frequency of Earths. If NASA decided to fly TPF-C instead of TPF-I, what would that mean for the collaborative effort with ESA's Darwin Mission? No problem. As ESA astronomer Malcolm Fridlund put it, "Darwinians are willing to evolve." ESA's astronomers wanted so badly to join NASA in the TPF effort that they would be willing to proceed with the best possible design for TPF, even if it was not the Darwin free-flyer.

Thus TPF not only survived the transition to a new NASA administrator, but now there were two TPFs to consider. Things were looking good for the search for living planets.

December 21, 2001—NASA headquarters announced that the Kepler Mission had been selected for flight as a Discovery-class mission, where missions are conceived and often led by non-NASA scientists who are called the mission's Principal Investigator. Twenty-six proposals had been submitted for the competition, and Kepler was one of the two winners. The other was the Dawn Mission to the two largest asteroids, Vesta and Ceres. The two missions were cost-capped at $299 million each, and both were scheduled for launch in 2006. Discovery Missions are used primarily to explore the Solar System, so NASA's willingness to use a precious Discovery slot to search for habitable worlds was encouraging. It was probably attributable in no small part to the enthusiasm for exoplanet searching of Edward Weiler, the head of NASA's Office of Space Science. My Carnegie colleague and Geophysical Laboratory director Wesley Huntress, a former head of the Office of Space Science himself, whose own Discovery Mission proposal had just been rejected, told me that he thought Kepler was a "shoo-in" in the Discovery Mission race because of Weiler's strong support.

Kepler was planned as a space telescope with a diameter of 38 inches (0.95 meter) designed to photograph a wide area of the sky in order to simultaneously monitor 100,000 stars for transits caused by their Earths. Kepler would stare at those 100,000 stars for 4 years, long enough for an Earth-like planet to transit its star at least three times. Three transits were needed to be certain of the planet's existence, because the transits caused by an Earth-sized planet are 100 times dimmer than those caused by a Jupiter-sized planet. Although hot Jupiters could be discovered with a 4-inch telescope in a parking lot in Boulder, finding Earths meant a space telescope like Kepler was needed to get above our Earth's atmosphere. Even then, one still had to contend with the turmoil and sunspots on the target star's surface, which could easily produce a dimming comparable to that of a transiting Earth.

The first of the transits told you nothing; it might be merely an artifact of something weird on the star's surface. The second transit told you the orbital period of the planet, if there really was a planet involved. This allowed you to predict when the third transit would occur, one orbital period later. If this third transit occurred as predicted, voilà! A new planet had been discovered, at least assuming that the Kepler team could rule out all other sources of false positives, such as periodic eclipses of background binary stars that were so faint they could not be seen next to the target star but whose light would insidiously blend in with the target star's light and mimic the dimming caused by an Earth-like transit.

Kepler's aperture of 38 inches was more than three times larger than CoRoT's 12-inch (0.3-meter) diameter, meaning that it had 10 times the area of CoRoT and so could monitor stars that were 10 times fainter. Kepler planned to follow about 100,000 stars in a region of the sky 10 degrees on a side and centered on the constellation

FIGURE 15. Transit of Venus in front of the Sun on June 8, 2004. Venus is about to end the transit by moving off the right edge of the Sun's disk. Venus is nearly the same size as Earth. [Courtesy of Robert Traube (Northern Virginia Astronomy Club). Image taken with a Celestron C-11 Schmidt telescope.]

Cygnus. For comparison, the Moon has a diameter of ½ degree, so the Kepler field would be as wide as 20 Moons lined up across the sky. CoRoT, on the other hand, images a square about 3 degrees on a side, only half of which is used for transit searching, so CoRoT's search area is more than 20 times smaller than Kepler's. Combined with the fact that Kepler planned to stare at the same stars for 4 years without ever blinking, and CoRoT planned to stare for no longer than 150 days, it was clear that Kepler should be able to outperform CoRoT on every count. CoRoT might be able to find "hot Earths" on

Mercury-like, short period orbits, but only Kepler would have the time to find habitable Earths with 1-year orbital periods around stars like the Sun.

If Kepler could jump through all the hoops of the NASA design and development process, get built correctly, and launch into an Earth-trailing orbit around the Sun rather than into the Atlantic Ocean off Cape Canaveral, Kepler was likely to beat the pants off CoRoT at finding true Earths.

January 30, 2003—Maciej Konacki of Caltech and his colleagues published a paper in *Nature* reporting the first detection of a planet by the transit method. HD 209458 b had been discovered by Doppler spectroscopy and then confirmed by transit photometry. Konacki managed to reverse this sequence for the first time.

A group of Polish astronomers from Warsaw University had built a 52-inch (1.3-meter) telescope at Carnegie's Las Campanas Observatory in Chile and were using it to study the myriad of stars toward the center of our Galaxy, in the so-called Galactic bulge. They were searching for microlensing events, where an unseen foreground star would pass in front of a visible Galactic bulge star. As predicted by Albert Einstein in 1936, the foreground star's gravity could bend the light coming from the visible star, causing the light rays to be concentrated in the direction of the Earth. The visible star would thus brighten for a few weeks and then return to normal as if nothing had happened. Einstein had predicted this gravitational lensing effect but did not believe that it could ever be observed. The Polish astronomers decided to prove him wrong. They named their search the Optical Gravitational Lensing Experiment, or OGLE, an appropriate acronym for a project that involves prolonged staring.

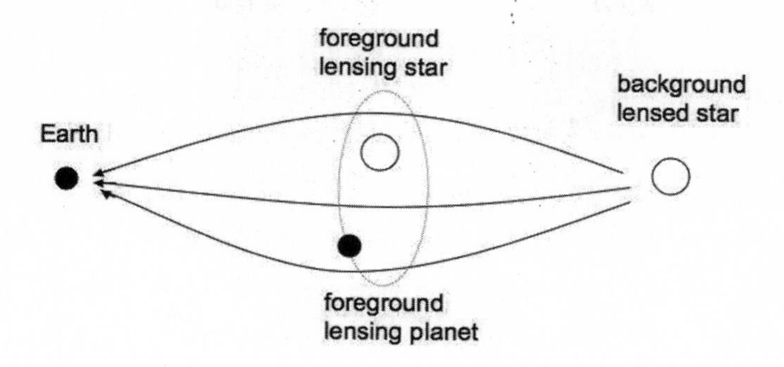

FIGURE 16. During a gravitational microlensing event, light from a background star is bent toward Earth by the gravitational pull from an unseen foreground star and planet, resulting in an apparent brightening of the background star for a period as long as a few months. The event occurs only once, because the foreground star continues on its orbit through the Galaxy and leaves behind the favorable alignment that by chance led to the event.

The OGLE team had been following 155,000 faint stars to look for microlensing events. As a byproduct of that search, they stumbled on a number of stars that periodically dimmed as if they had transiting planets. In 2002, Andrzej Udalski of the Warsaw University Observatory had published a list of 137 stars that might have transiting planets. Given the likelihood that most of these candidates were false positives caused by things like unseen eclipsing binary stars behind the target star, Udalski left it to others to follow up on the list and see whether any of them could be proved by Doppler spectroscopy to have planets. Konacki did the dirty work and found the first one, named OGLE-TR-56, the 56th star on Udalski's list.

OGLE-TR-56 was orbited by a planet with a mass of 0.9 Jupiter mass, which sounded reasonable, but it was orbiting with a period of only 1.2 days, much less than 51 Pegasi b's 4.2 days or HD 209458 b's 3.5 days. The Doppler planet search folks had never seen such a short-period planet in their surveys, and they wondered why not. Konacki had measured the Doppler wobble with a total of only three data points, so few as to make the claimed detection laughable from the point of view of the Doppler teams, who routinely take dozens of measurements before daring to claim a detection. If it was real, one thing was clear: because its 1.2-day period meant that it was closer to its star than any other planet, OGLE-TR-56 b was likely to be the hottest Jupiter of them all. Real estate agents would not want listings for this hellish world.

April 30, 2003—Artie Hatzes of the Tautenberg Observatory in Germany, Gordon Walker, and their colleagues submitted a paper to the *Astrophysical Journal* providing good evidence that Gamma Cephei had a giant planet after all, in accordance with the claim made by Bruce Campbell in 1988 only to be withdrawn by Walker in 1992. A total of 21 years of Doppler data on the Gamma Cephei binary star system implied that it contained a planet with a mass of at least 1.7 Jupiter masses that orbited Gamma Cephei A (the brighter star) every 2.5 years. This planet was the same one that the Canadian group had agonized over in 1988–1992. The new decade of data ruled out the possibility that the 2.5-year variations were caused by changes in the stellar atmosphere, the troubling worry that had caused the retraction in 1992. The Canadians had been denied the glory of the discovery of the first extrasolar planet in 1988, well before the pulsar planets arrived in 1992 and 51 Pegasi b appeared in 1995—robbed of this unique distinction by their own caution and uncertainty.

June 30, 2003—The Canadians beat the French in the race to get the first stellar seismology telescope into space. The Microvariability and Oscillations of Stars Telescope (MOST) microsatellite was launched on a former Soviet intercontinental ballistic missile (ICBM) from Plesetsk, the primary launch site for Soviet and Russian military satellites. MOST was dubbed Canada's "Humble Space Telescope" because of its small size, about that of a suitcase, and its small cost, about $10 million Canadian. Now that transiting planets were beginning to be discovered, though, it was clear that MOST could also monitor stars like HD 209458 and learn more about their hot Jupiters, all for a bargain basement price. Canada's Gordon Walker was back in the extrasolar planet game, this time looking for transits instead of those vexing Doppler wobbles.

CHAPTER 4

The Mars Gold Rush

There is life on Mars, and it is us.

—RAY BRADBURY (JULY 1976)

January 14, 2004—President George W. Bush addressed the nation from the auditorium at NASA headquarters. Bush presented his new Vision for Space Exploration, intended to provide NASA with detailed guidance on exactly what it should be doing in the painful aftermath of the Columbia disaster, when a Space Shuttle disintegrated during its return from space on February 1, 2003. Shuttle flights had been suspended since then, which halted work on the International Space Station and delayed the plan to service and update Hubble one last time. Columbia had been scheduled to visit Hubble early in 2005.

Bush took charge of NASA's quandary and laid out a new game plan. The prioritized list for NASA was clear-cut: first, return the Shuttle to flight; second, finish building the Space Station; third, prepare to retire the Shuttle by developing a new space transportation system; and fourth, return human beings to the Moon while on the way to Mars.

Mars? Mars?? Bush's plan was breathtaking in its scope even without astronauts heading off to Mars. Evidently President Bush was attempting to outdo his father, the first President Bush, with the Vision for Space Exploration. The first President Bush had proposed around 1989 to send NASA astronauts to Mars for about $400 billion, a price tag that outraged Congress and led to the demise of the first Bush's Mars initiative. The first President Bush had kicked Saddam Hussein out of Kuwait after Hussein invaded it but had left Hussein as the ruler of Iraq. The second President Bush seemed to have decided to address both of these paternal challenges, first with the U.S. invasion of Iraq in 2003, and now with the Vision for Space Exploration. The fact that deposing Saddam Hussein had turned into a disaster of unprecedented proportions for the Iraqi people did not bode well for the success of the new Mars initiative: the Bush administration did not have a good track record for knowing what to do after the missiles were launched.

The Jet Propulsion Laboratory (JPL) at Pasadena, California, had landed the first Viking spacecraft on Mars on July 20, 1976 (exactly 7 years after Apollo astronaut Neil Armstrong set foot on the Moon), thus beginning a search for life on Mars that went well beyond looking for canals and interstates. In the 30-odd years since Viking, astrobiologists had realized that, rather than searching directly for Martian microbes, as Viking had done, it made more sense to look first for habitable regions on Mars, just as astronomers were planning to do in the search for habitable planets. In both cases, the key ingredient is water, the essence of terrestrial life. NASA's mantra for looking for life on Mars became "follow the water"—that is, find regions where water was present now or had been present in the past on Mars. Once evidence for liquid water was found, that would be the time to start the search for Martian bugs.

By 2004, the possibility that the ALH84001 meteorite had proved that there was life on Mars had been discounted by nearly every savvy astrobiologist except those who had made the initial claims, David McKay and Everett Gibson. With the advent of President Bush's Vision, the exploration of Mars became a driving force behind not only NASA's science goals but also the future of human space flight. The search for Martian water would begin with robotic missions, though, and perhaps the robots would find some good places for astronauts to explore several decades later. If robots or astronauts could discover life on Mars, life that arose independently of life on Earth, the implications were enormous for life elsewhere in the Galaxy and the universe. Widespread life would be inevitable, at least if habitable planets were common.

Almost as an afterthought, and apparently thanks to the advice of NASA administrator Sean O'Keefe, the Vision for Space Exploration included a call, buried deep within the Executive Summary in a section entitled "Mars and Other Destinations," for "advanced telescopic searches" for habitable worlds. Its presence was cause for giddy celebration among the planet hunters, who knew the importance of this single phrase: the Bush administration was now committed to pushing NASA forward with its grand plans to image other worlds outside our Solar System. The Vision seemed to be proof that NASA now had a vision worthy of its past accomplishments and its future potential. All NASA headquarters needed was the money to carry out the Bush plan.

January 16, 2004—Steven Beckwith, director of the Space Telescope Science Institute, sent out an e-mail to his staff telling them that he had just had an unsettling meeting with Sean O'Keefe.

O'Keefe told Beckwith that he had decided to cancel the upcoming Shuttle mission to Hubble. This mission was known as Servicing Mission 4, or SM4, although it was really the fifth mission, because an extra mission had had to be inserted in the queue to bring Hubble back to life after the loss of the fourth of its six gyroscopes in 1999 (without three gyroscopes, Hubble could not point itself correctly in space). The cancellation of SM4 meant that Hubble's days were numbered; SM4 was to have installed six fresh new gyroscopes and replaced all six of the original batteries, which had been launched with Hubble in 1990. Even the Energizer Bunny slows down after a couple of decades of constant use in the harsh environment of space.

O'Keefe said the decision to cancel SM4 was his decision and his alone. The reason lay in the fact that Hubble was in a low-Earth orbit unlike the low-Earth orbit of the International Space Station. Hubble's orbit is inclined about 20 degrees closer to Earth's equator than the Space Station's orbit. Once it was launched toward Hubble, a Shuttle could not change its orbit enough to make it to the International Space Station in case of trouble. NASA had determined that the loss of Columbia was caused by falling foam insulation during launch, which had impacted the leading edge of Columbia's left wing, punching an unseen hole that would prove deadly upon return to Earth. NASA had been working hard to fix this problem, but for safety's sake O'Keefe decided that all future Shuttle flights must be to the International Space Station, where the crew could find a safe haven and wait for rescue in the event that their Shuttle sustained potentially fatal damage during launch.

By 2004, Hubble was down to only three functioning gyroscopes. The expectation was that another gyroscope might fail by 2005, and yet another by 2007. At that point, Hubble would be as good as dead. Without SM4, the Hubble Era was nearly over.

Hubble would then have no overlap with the Next Generation Space Telescope, which was now called the James Webb Space Telescope in honor of the early NASA administrator who insisted that NASA balance human space flight with a healthy program of scientific investigation. The Webb Telescope was now planned for launch in 2011, and its launch might slip further. The plan was that the Space Telescope Science Institute would continue to be supported in the interim before the launch of the Webb Telescope gave everybody something to do again. Even astronomers outside of the Institute could expect to survive the severe drought between Hubble and Webb, by availing themselves of NASA grants to use Hubble's extensive archives of images and spectra in their continuing researches. Sean O'Keefe thought his plan made good—though painful—sense, given the pickle in which NASA found itself.

February 12, 2004—Ian Bond of the Institute for Astronomy in Edinburgh, Scotland, Andrzej Udalski, and their colleagues submitted a paper to the *Astrophysical Journal* Letters claiming the first robust detection of an extrasolar planet by the gravitational microlensing method. Einstein would have been amazed that an idea he had so easily dismissed as impractical could be put to such good use. Bond and Udalski's planet was given the unwieldy name of OGLE 2003-BLG-235/MOA 2003-BLG-53 b, because it involved observations taken by two different groups searching for microlensing events: Udalski's OGLE team and Bond's Microlensing Observations in Astrophysics (MOA) team. A total of 32 astronomers had their names on this historic paper. Compared to the Doppler or transit planet searches, which involved handfuls of people, microlensing required large teams of astronomers akin to those on space-based telescopes or to the large groups of physicists working at high-energy particle accelerators.

The need for large teams arose from the peculiar nature of a microlensing event caused by a planet. When an unseen foreground star passes in front of a visible star, the resulting brightening of the visible star can last for a period of several months. If the foreground star happens to have a planet in orbit around it, and if the planet's position is aligned in the right way, the planet's gravity can add to the foreground star's gravity to further deflect the light from the visible star in the direction of Earth. The resulting signature of a planet around the foreground star is then a spike of increased brightness of the visible star, which may lead to considerably more brightening than that caused by the foreground star alone. However, the spike due to the planet is of short duration; it typically lasts only a few hours or less.

Detecting a planet by microlensing thus requires obsessive attention for several months: once an event starts, the star must be monitored continuously in hopes of catching the brief spikes of brightening caused by a planet. Given the vagaries of weather, with occasional cloudy nights affecting even the most cloud-free observatory sites, random instrument problems, and the absolute certainty that the Sun will rise at dawn each day, microlensing observations require a network of telescopes located all around the world, at three or more different longitudes in both the Northern and Southern Hemispheres. The MOA and OGLE international collaborations were thus essential to making progress in this endeavor.

Bond and Udalski's OGLE 2003-BLG-235/MOA 2003-BLG-53 b planet appeared to be a gas giant with a mass of 1.5 Jupiter masses, orbiting at three times the Earth–Sun distance, equivalent to the asteroid belt in our Solar System. The star it orbited could not be seen, but given the statistics of stars in our Galaxy, the best bet was that

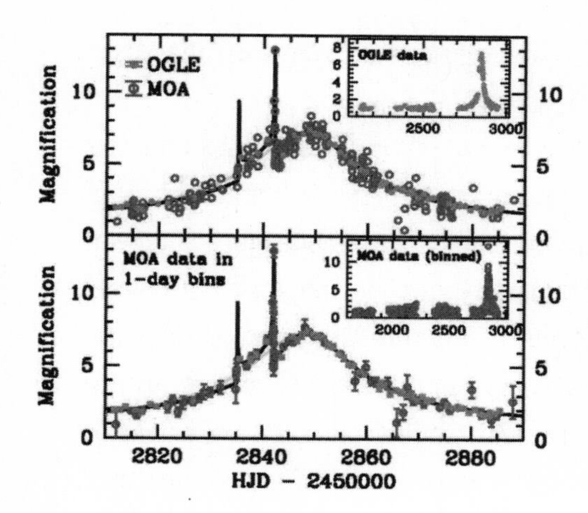

FIGURE 17. First detection of a planet by the gravitational microlensing technique. The two spikes in brightness of the background star are caused by a planet orbiting the foreground star, which results in the overall rise and fall in brightness over a time period of several months. [Reprinted, by permission, from I. Bond et al., 2004, *ApJ*, volume 606, page L156. Copyright 2004 by the American Astronomical Society.]

the planet orbited an M dwarf star. M dwarfs are faint stars even when nearby, and at the distance of the Galactic bulge, roughly 20,000 light-years away, an M dwarf would be much harder to see.

There had been a prediction back in 1993 that microlensing would reveal exoplanets faster than any other ground-based technique—an optimistic claim, given that the first such detection had taken 10 years. But microlensing did have unique promise: because of the pathological bending of light associated with Einstein's effect, even a planet as small in mass as Mars could produce a detectable spike. Such events might be rare, but they would be detectable. In order to guarantee the sort of round-the-clock observations that were necessary to catch planets in the act, David Bennett of the University of Notre Dame, Indiana, had proposed building a space telescope dedicated to microlensing and named the Microlensing Planet Finder, or MPF. Like Kepler, MPF would stare at a field of stars, although it would look toward the Galactic center instead of Cygnus. Like Kepler, MPF had been proposed as a Discovery Mission. Unlike Kepler, MPF had not been selected, but Bennett was a patient man, and the fact that microlensing finally had some points on the board could only improve the chances for MPF's selection some day.

Now there were four proven methods for discovering extrasolar planets: pulsar timing, Doppler shifts, transits, and microlensing. Fritz Benedict of the University of Texas and his colleagues had been able to detect the first M dwarf planet found by Doppler spectroscopy (Gliese 876 b) by using Hubble to measure the star's astrometric wobble. But Hubble time was too precious to allow anyone to use Hubble to undertake a blind search for new planets by astrometry. Hubble astrometry had to be reserved for following up rather than searching. SIM would do the astrometric planet search that Hubble could not do, and SIM would do it nearly a thousand times better. Although Hubble could detect the relatively large wobble caused by a gas giant around an M dwarf, only SIM could search for Earths around Sun-like stars.

April 12, 2004—TPF Project Scientist Chas Beichman announced that NASA and JPL had changed their plans. There would no longer be a decision in 2006 to continue to develop only one option for TPF, either the optical coronagraph or the infrared interferometer. Heartened by the boost for extrasolar planet searches laid out for NASA in President Bush's Vision for Space Exploration, NASA and JPL now planned to fly both versions of TPF. TPF-C was planned for launch in 2014, followed by TPF-I in 2016. The case for life on a nearby world would be much stronger if biomarkers could be seen at both visible and infrared wavelengths. Malcolm Fridlund chimed in that the European Space Agency was pleased by the decision and that ESA still hoped to be able to combine its Darwin free-flyer concept with NASA's TPF-I in a joint mission.

July 2, 2004—*Science* reported that Sean O'Keefe had undertaken the most drastic organizational change in over a decade of NASA history. The new organization chart bore little resemblance to the one from the day before. Ed Weiler had been plucked out of NASA headquarters to become the head of NASA's Goddard Space Flight Center, replacing Alphonso Diaz, who was given Weiler's previous job. Weiler would now be more directly responsible for Hubble and Webb—and less involved with SIM and the TPFs. The loss of wily Ed Weiler to Goddard did not auger well for planet hunting.

The "three envelopes" school of management holds that whenever a new NASA administrator is appointed, he is handed three numbered envelopes by the departing administrator and told to open them one-by-one as the inevitable crises arise. The first envelope holds a message saying, "Blame your predecessors." O'Keefe had suffered through the Columbia tragedy, which could plausibly be blamed on the failures of previous administrators to correct a known problem with the Shuttle's foam insulation. The message in the second envelope says simply, "Reorganize." O'Keefe was doing that now. The third envelope's message reads, "Prepare three envelopes." O'Keefe had one more to open.

July 12, 2004—Paul Butler and his colleagues submitted to the *Astrophysical Journal* a paper about the Doppler detection of the first example of what was probably a new class of planets. Butler's new planet orbited the the star Gliese 436, an M dwarf star with a mass less than half that of the Sun. This was only the second M dwarf star with a planet found by the Doppler method; the rest of the more than 100 planets that had been found by Doppler searches orbited Sun-like stars.

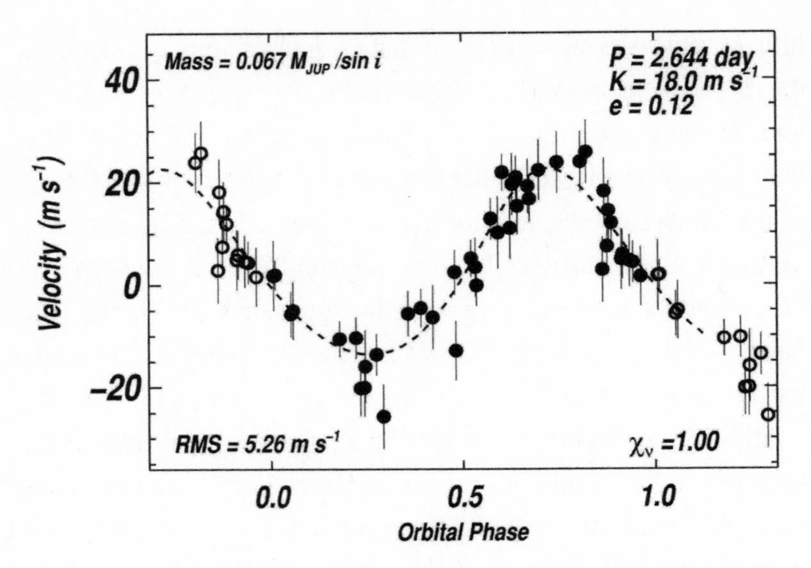

FIGURE 18. First detection of a hot super-Earth by the Doppler wobble technique. This planet is in orbit around the M dwarf star GJ 436 and has a minimum mass about 21 times that of Earth. [Reprinted, by permission, from R. P. Butler et al., 2004, *ApJ*, volume 617, page 584. Copyright 2004 by the American Astronomical Society.]

But the most amazing fact about the new planet was its extraordinarily low mass: it seemed to be only slightly more massive than Neptune, with a minimum mass of 21 Earth masses. The lowest-mass planet previously known had also been found by Butler and colleagues; it orbited the star HD 49674 and had a minimum mass about two times higher than that of Gliese 436 b. Considering that Saturn's mass is 95 Earth masses, the HD 49674 planet might be viewed as a pint-sized version of the gas giant Saturn, but Gliese 436 b apparently occupied a different territory: the land of ice giants, like Neptune and Uranus.

What was Gliese 436 b made of? Was it really an ice giant, or something else? One clue was that it completed its orbit in only 2.6

days, implying an orbital separation from the M dwarf Gliese 436 of 0.028 AU. Although Gliese 436 was an M dwarf, not nearly as luminous a star as the Sun, Gliese 436 b nevertheless had to be toasty at that distance. An icy world might not stay icy long on such an orbit, so maybe Gliese 436 b was a water world, or maybe it was just a gigantic rock—the first extrasolar terrestrial planet, a hot super-Earth. Whatever it was, Gliese 436 b seemed to be the first member of a new class of exoplanets: pulsar planets, gas giants, and now super-Earths, whatever *they* were.

More were to follow shortly. Less than a month later, Barbara McArthur of the University of Texas and her colleagues, including Butler and Marcy, submitted a paper to the *Astrophysical Journal* reporting another super-Earth. The solar-type star 55 Cancri had already been known through Doppler spectroscopy to have three gas giant planets, nicknamed 55 Cnc b, 55 Cnc c, and 55 Cnc d. McArthur and her team added a fourth, setting two records at once. 55 Cancri had more known exoplanets than any other star, and it had the honor of hosting the lowest-mass planet. The minimum mass of 55 Cnc e was 14 Earth masses, well below the 21 Earth masses of Gliese 436 b. The newcomer was also on a short, 2.8-day orbit, making it the second hot super-Earth.

If hot super-Earths existed, then the ongoing, patient, deliberate Doppler searches could be expected to find super-Earths on longer-period orbits: warm super-Earths with liquid water on their surfaces. Even so, the strong gravity on the surface of a warm super-Earth might make taking an afternoon stroll a bit tiresome for beings visiting from Earth.

CHAPTER 5

Instability: If I Did It, Here's How It Happened

A few Christmases ago, a science writer asked me: With regard to your research, if Santa Claus agreed to bring you the answer to any question, what would it be? I told her that I would like to find out how Jupiter formed.

—GEORGE W. WETHERILL (DECEMBER 16, 1997)

July 26, 2004—The Second TPF/Darwin International Conference began at Mission Bay in San Diego, California, a year after the first joint meeting was held in Heidelberg, Germany. One of the highlights of the Mission Bay meeting was to be a great debate on the first day about the formation of giant planets, not only the four in our Solar System—Jupiter, Saturn, Uranus, and Neptune—but also those found around other stars, which now numbered well over 100. With that many examples of giant planets, it seemed timely to summarize what was known about how giant planets formed, and what their formation meant for the formation of Earth-like worlds.

Giant planets such as Jupiter are second only to their stars in terms of power and dominance in planetary systems. Jupiter has 318 times the mass of Earth, and Saturn has 95 times Earth's mass, making the terrestrial planets—Mercury, Venus, Earth, and Mars—seem like inconsequential also-rans in the planet formation race. In some ways, that assessment is true; planet formation theorists believe that the terrestrial planets took tens of millions of years to grow close to their present sizes, whereas the gas giants must have formed within a few million years or so. Most of the mass contained in gas giant planets is hydrogen and helium gas, with only a minor fraction coming from the iron and silicate rocks that form the bulk of the terrestrial planets. These rocks can take their time in forming a planet such as Earth, but the gas needed for forming gas giant planets is known to disappear from most young stars on time scales of a few million years or less. Hence, if a gas giant planet is going to form at all, it must do so before the planet-forming disk's gas disappears through a combination of being eaten by the voracious young protostar and being blown back out into space by winds and radiation from the protostar or other nearby young stars. The fact that giant planets must have crossed the finish line early, while the terrestrial planets were still trying to get out of the starting gate, meant that any theorist who wanted to understand how Earth formed needed to know how Jupiter formed. With such a large mass, Jupiter's periodic gravitational pulls and yanks would have had a major effect on the planetary embryos that were trying to find each other on the Dodge-Em car ride.

The quotation at the beginning of this chapter was uttered by George Wetherill on the occasion of his receipt, from President Bill Clinton in 1997, of the National Medal of Science, the nation's highest scientific award. Wetherill had received this medal in large part for his pioneering work on understanding how the Earth and other terrestrial planets formed from a rotating disk of gas and dust particles. But Wetherill knew that before he could understand how Earth formed, he needed to understand Jupiter's formation. In the absence of that God-like knowledge, all Wetherill could do was guess and then consider the consequences for the making of Earth if Jupiter had formed this way, or if it had formed that way.

Terrestrial planets might be insignificant in terms of the dominance of planetary systems, but they appear to be critical as abodes for life, so without a good theoretical understanding of how they form, it is impossible to predict how many living planets await us in the Galaxy. The great debate at the Mission Bay conference was intended to answer Wetherill's question through dialogue between proponents of the two competing theories of giant planet formation. There are only two basic ways in which a gas giant planet can be made: from the bottom up, or from the top down.

The bottom-up method relies on the same basic process of planet building that Wetherill and every other reasonable theorist believe led to the formation of the terrestrial planets: banging together enough solid bodies in the Dodge-Em car ride to make a planet. The key difference for giant planet formation was that the Dodge-Em car ride had to lead to the formation of bodies roughly ten times the mass of Earth before the ride was over and the disk gas disappeared from the fairgrounds, like a heavy fog lifting after sunrise. If a planetary embryo of 10 Earth masses formed fast enough, this body might be able to pull in hydrogen and helium gas from the planet-forming disk, gathering a dense atmosphere around the solid core of the growing protoplanet. The added gas mass would make the planet's gravity even stronger, enabling it to pull more and more gas onto the core—a process called accretion. Once gas accretion started, it was thought to be a runaway process, so that a 10-Earth-mass core could

FIGURE 19. Sketch of Douglas Lin, one of the leading planet formation theorists, at the time of the 1985 NASA Ames nebula workshop. [Sketch by and courtesy of Patrick Cassen.]

quickly grow to become a 318-Earth-mass gas giant planet like Jupiter.

This process of *core accretion* is the generally favored mechanism for gas giant planet formation, in large part because it is a natural extension of the same planet-building process that all acknowledge must occur when terrestrial planets form. The meeting organizers asked Douglas Lin of UC Santa Cruz, a prolific planet formation theorist well acquainted with core accretion, to present the case for the bottom-up mechanism at the Mission Bay meeting.

Proponents of the top-down mechanism, by contrast, envision clumps of gas and dust forming directly out of the planet-forming disk as a result of the self-gravity of the disk gas. The clumps would

FIGURE 20. Sketch of Alastair Cameron [1925–2005], another leading planet formation theorist, at the time of the 1985 NASA Ames nebula workshop. [Sketch by and courtesy of Patrick Cassen.]

result from the intersections of random waves sloshing around the disk, waves that look much like the arms in spiral galaxies such as the Milky Way. When two spiral arms pass through each other, they momentarily merge to form a wave with their combined heights, just as waves do on the surface of an ocean. Such a rogue wave might rapidly lead to the formation of a clump massive enough to be self-gravitating and so hold itself together against the forces trying to pull it apart. Once such a self-gravitating clump forms, the dust grains within the clumps settle down to the center of the protoplanet and form a core, leading to the same basic structure (solid core and gaseous envelope) produced by the core accretion mechanism. This top-down mechanism is termed *disk instability*. In comparison to

FIGURE 21. Victor Safronov [1917–1999], the Soviet pioneer in developing the basic theory of Earth's formation from a population of smaller bodies. [Courtesy of V. S. Safronov.]

core accretion, disk instability is considered the dark horse candidate because it requires new physics. Rather than relying solely on collisions between rocks, disk instability invokes gravitational instabilities in planet-forming disks and requires these instabilities to be able to form self-gravitating clumps rapidly. Few planet formation theorists are willing to bet on disk instability; nearly all stick with core accretion when it is time to lay their money down.

Disk instability had been proposed as a gas giant planet formation mechanism in 1978 by Alastair Cameron of Harvard University. Cameron had also proposed core accretion in 1972, although both ideas had antecedents in the (independent) work of Gerard Kuiper of the University of Chicago and Victor Safronov of Moscow's Schmidt Institute. Al Cameron was used to coming up with new ideas and abandoning his older ones. By 1982 Cameron had tired of his 1978 idea of "giant gaseous protoplanets" and discarded it. I revived the idea in a *Science* paper in 1997 and provided detailed computer models showing how a gravitational instability could lead to the formation of clumps. The Mission Bay meeting organizers asked me to present the heretical case for disk instability.

The great debate turned out to be two talks on the same subject, rather than a true back-and-forth between the advocates of core accretion and those of disk instability. Doug Lin gave his talk, I gave mine, and then there was a period for questions from the audience.

My impression was that we both had good points to make about how giant planets could form.

The celebrated UC Berkeley theorist Frank Shu, then president of the National Tsinghua University in Taiwan, was asked to score the debate in his concluding remarks at the conference. Shu scored it as a toss-up. Maybe both processes were at work in our Solar System, he said, with disk instability first forming Jupiter, and then core accretion taking over to form Saturn, Uranus, and Neptune. Compared to previous dour assessments of the chances for disk instability to produce gas giant planets, this was a seismic shift. Disk instability was now at least a possibility, rather than an idea that could be dismissed as impossible or implausible.

Core accretion is an intrinsically slow process, because the gas accretion phase must await the growth of a sizable solid core before things really get going. One way to speed things up is to consider a more massive planet-forming disk, with more dust particles to serve as the building blocks for the growing planetary embryos. Satoshi Inaba had worked with George Wetherill at the Department of Terrestrial Magnetism on trying to grow the massive cores fast enough for there still to be some disk gas around for the gas accretion phase. They had reported, in a 2003 *Icarus* paper, that Jupiter's core could have grown in less than 4 million years—barely fast enough, but only if they assumed that the planet-forming disk had a mass on the order of 10% of the mass of the young Sun.

A disk with as much mass as Inaba and Wetherill needed to form Jupiter's core was likely to be gravitationally unstable. Any small density enhancement would tend to pull more disk gas toward it, because its gravitational attraction would stand out from that of the smoother portions of the disk. As the enhancement gained mass, its gravitational attraction would further increase, leading to the inward flow of even more mass. This would be a runaway process like that invoked in the gas accretion phase of core accretion, although both processes would be limited by forces acting in the opposite direction, such as the thermal pressure of the disk gas, and the rotation of the disk gas as it tries to find a way to flow onto a core or a clump. The big advantage of disk instability was that it could proceed directly to this phase of runaway growth without having to wait for a massive core to build up first via the banging together of increasingly larger planetesimals. The first step of core accretion might be so slow that the disk gas would be gone by the time most cores grew large enough to accrete the gas, resulting not in gas giants but in "failed cores." Disk instability does not have this problem. If disk instability worked, extrasolar Jupiters would be the rule rather than the exception, so most extrasolar Earths would have their Jupiters in place, ready to protect their life forms from wayward comets, as Wetherill had required in 1992. Models of terrestrial planet formation published in *Science* in 2001 by the Department of Terrestrial Magnetism's Stephen Kortenkamp, Wetherill, and Inaba, had shown that having Jupiter form quickly by disk instability could actually encourage the formation of habitable worlds. My colleague John Chamber's 2003 models in particular had suggested that the number of habitable worlds might be increased by 50% if Jupiter were not a slowpoke in forming. Disk instability seems to be not just compatible with, but also highly supportive of, the formation of habitable worlds.

Disk instability operates on a time scale of a thousand years or less, whereas core accretion requires a million years or more. If Inaba and Wetherill needed a gravitationally unstable disk to get core accretion started, even a disk that was just marginally gravitationally unstable, it was likely that disk instability would be able to run laps around core accretion on such a race track. Of course, if the disk's mass was too small for it to be gravitationally unstable, disk instability could never get out of the starting gate, and core accretion could take its time getting around the track.

At the Mission Bay meeting I made the point that disk instability was so fast that it could form gas giant planets even in the shortest-lived disks. Hubble had been used to image planet-forming disks, known as protoplanetary disks or "proplyds," in regions where most stars form, namely regions of the Milky Way where thousands of stars or more form together in massive clusters, with the stellar masses ranging from lowly M dwarfs to the mightiest O stars. Astronomers had found that such disks in star-forming regions like the Orion Nebula Cluster and the Eta Carina Nebula were bathed in ultraviolet light emanating from the massive stars in their vicinities. The ultraviolet light was strong enough to heat the normally cold hydrogen gas in the disks to temperatures of many thousands of degrees—hot enough for the gas to flow away from the disks and return to space. Astronomers estimated that the disk gas could be lost in less than a million years. Because core accretion could not form a gas giant planet that rapidly, there was an uneasy feeling arising that maybe gas giants were rare beasts, and the Solar System an oddity, because most stars could not be expected to have disks that would last long enough for core accretion to do its job.

If disk instability could make gas giants in even the shortest-lived proplyd, there would be no problem with making Jupiters practically anywhere the disks were massive enough. I was writing an e-mail about this reassuring thought to the *Washington Post*'s science reporter, Kathy Sawyer, in 2001 when a lightbulb went on over my head. It dawned on me that if disk instability had made gas giant protoplanets in a gaseous disk being evaporated by the ultraviolet light from a nearby massive star, then once the outer disk gas was gone, the outer protoplanets would be the next to be subjected to the harsh attentions of the O star. The protoplanets would lose their hydrogen gas envelopes because of the O star's ultraviolet light, just as the disk had earlier, and they would be stripped down to their central, solid cores of ice and rock, with only a minimal amount of gas left trapped in their gravitational grip. The end result would be conversion of the outer giant gaseous protoplanets into ice giants—that is, planets like Uranus and Neptune, which are composed primarily of rock and ice, with only a veneer of hydrogen, helium, and other gases. The process would not affect the innermost gas giants, because the star's gravitational pull would be strong enough to keep the hot gas from escaping inside some critical distance from the star. This critical distance is roughly the same as Saturn's distance for a solar-mass star.

The implications were breathtaking. This new scenario implied that there was a good chance that our own Solar System had formed in a region where stars formed in huge clusters along with massive stars. The ultraviolet light from the massive stars could have converted the outermost planets from gas giants into Uranus and Neptune, whereas Jupiter would complete its evolution into a gas giant, unmolested by ultraviolet radiation. Saturn would be an intermediate case, losing some of its gaseous envelope but not all of it, which would explain why its mass was less than one-third that of Jupiter's. Earth and the other terrestrial planets would be oblivious to the ultraviolet photoevaporation that was stripping the gas out of the outer disk and planets.

If correct, this scenario meant that the Solar System might have formed in the same type of Galactic locales where most stars formed. By implication, most stars in the Milky Way might harbor planetary systems similar to our own. The estimated number of habitable worlds might increase by a factor of 10. Suddenly the universe was looking to be a considerably more crowded place than it was before I answered Kathy Sawyer's e-mail.

Wetherill and I and our Department of Terrestrial Magnetism colleague Nader Haghighipour published this intoxicating idea in *Icarus* in 2003. The implications were important for the optimal design for TPF/Darwin: if Earths really were 10 times more commonplace than might be the case otherwise, TPF/Darwin could be designed to search a much smaller number of closer stars and still detect and characterize a goodly number of habitable worlds. This would make TPF/Darwin a lot cheaper to build, and cost always looms as a potential showstopper for any space mission.

August 9, 2004—Sean O'Keefe bowed to pressure to keep Hubble running and said that NASA would ask Congress for the money to launch a robotic servicing mission that would not place astronaut lives in jeopardy. O'Keefe's decision to drop SM4 had been met with widespread dismay in the astronomical community that relies on Hubble, and among a public that had come to rely on Hubble for fresh new photos of the heavens. A robotic servicing mission was expected to cost at least $2 billion; just bringing Hubble back down to Earth with a robot might cost $400 million. But where would the money come from to save Hubble, given JPL's plans to build SIM, TPF-C, and TPF-I? All three missions had been combined into the "Navigator Program" at JPL, and Navigator evidently needed big bucks to fulfill its promise of finding Earths.

August 25, 2004—Portuguese astronomer Nuno Santos and his Swiss–French colleagues submitted a paper to *Astronomy & Astrophysics* claiming discovery of a third planet in the Mu Ara system, with a minimum mass of 14 Earth masses and an orbital period of 9.5 days: a third hot super-Earth had been found by Doppler spectroscopy.

December 9, 2004—An ad hoc National Academy of Sciences study group recommended that NASA plan on using a Shuttle flight to perform SM4, not the robotic mission that had been proposed by O'Keefe as a compromise.

December 13, 2004—Sean O'Keefe sent a letter to President Bush resigning as NASA administrator, effective as soon as a replacement was found. O'Keefe had been offered the position of chancellor of the Louisiana State University in Baton Rouge at a salary of $425,000 per year, several times his NASA salary. Evidently O'Keefe had opened the third envelope and decided it was time to cash in his chips and go home to Louisiana.

March 12, 2005—Michael Griffin was chosen to succeed Sean O'Keefe as NASA administrator. Griffin was a physicist and engineer at the Johns Hopkins Applied Physics Laboratory in Laurel, Maryland. A native of Aberdeen, Maryland, Griffin holds degrees from Johns Hopkins University in Baltimore and the University of Maryland. Not surprisingly, Maryland Senator Barbara Mikulski welcomed his nomination. Presumably O'Keefe prepared three new envelopes for his successor.

March 22, 2005—A press conference was held in the Webb Auditorium at NASA headquarters announcing the first detection of light

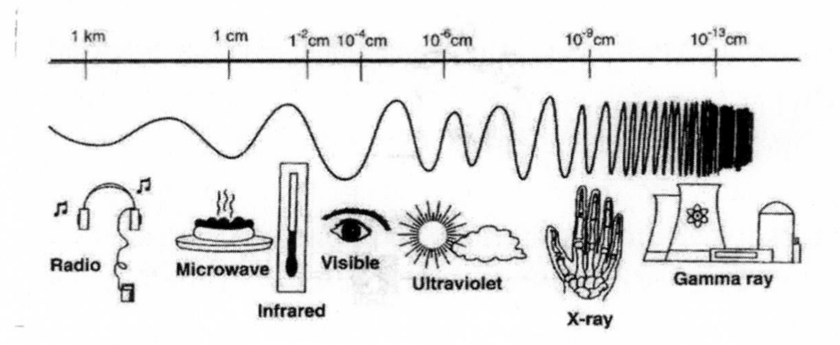

FIGURE 22. The spectrum of light at different wavelengths, ranging from long-wavelength radio waves to short-wavelength gamma rays. [Courtesy of NASA].

from an extrasolar planet. Two different teams, one headed by David Charbonneau, now an assistant professor at Harvard, the other by Drake Deming of NASA's Goddard Space Flight Center, had accomplished this feat using NASA's 34-inch (0.85-meter) Spitzer Space Telescope, named after Princeton University astronomer Lyman Spitzer, an early advocate of large space telescopes. Spitzer had been launched in August 2003 as the last of the four "Great Observatories" NASA built to cover four increasingly longer-wavelength bands of light: the Compton Gamma-Ray Observatory, the Chandrasekhar X-Ray Observatory, the Hubble Space Telescope, which covered ultraviolet to near-infrared wavelengths, and now Spitzer, a purely infrared space telescope.

The press conference had to be held one day early, when a British magazine broke the embargo on the story: Deming's paper on his team's Spitzer discovery was to be published in *Nature* on March 23. The story drew worldwide attention, with front page stories in the *New York Times* and other leading newspapers. The press realized that this story was a big one.

Spitzer's ability to capture infrared light with much longer wavelengths than what Hubble could see meant that Spitzer had a chance to detect light that was emitted by a hot Jupiter. Hot Jupiters could be seen in visible light, most of which is reflected light that originated on the star, but most of their own light is given off as infrared light at wavelengths that are several times longer than visible light. Spitzer was like Hubble on steroids, able to do things that Hubble could not, in spite of having only one-third its diameter. At 34 inches, Spitzer is the largest and most sensitive infrared astronomical telescope ever flown in space.

Both teams had used Spitzer to make the first direct detection of light from an extrasolar planet, but each team studied a different hot Jupiter. Deming's team worked on the first transiting planet, HD 209458 b, whereas Charbonneau's team observed the first transiting planet found by Tim Brown's Trans-Atlantic Exoplanet Survey, called TrES-1 b.

Transits occur when planets pass in front of their stars. *Secondary eclipse* is the name for what happens when the planet passes behind its star, such that the planet's light is blocked by its star. That means that the total amount of light from the star and the planet must decrease not only during a transit, or primary eclipse, but also during a secondary eclipse, when the planet's light cannot be seen. At infrared wavelengths, the loss of a hot Jupiter's light during the secondary eclipse results in a decrease of about 0.25%, small but detectable with Spitzer. At visible wavelengths, the dip during the secondary eclipse is about 25 times smaller than the infrared dip and so is not discernible, even by Hubble.

Spitzer was designed to be a general-purpose infrared telescope, not a planet hunter, but it had nevertheless managed to make the first direct detection of an extrasolar planet. Still, Spitzer's feat only whetted the appetite for what specially designed space telescopes such as TPF-C and TPF-I could do in the direct detection business: the TPF

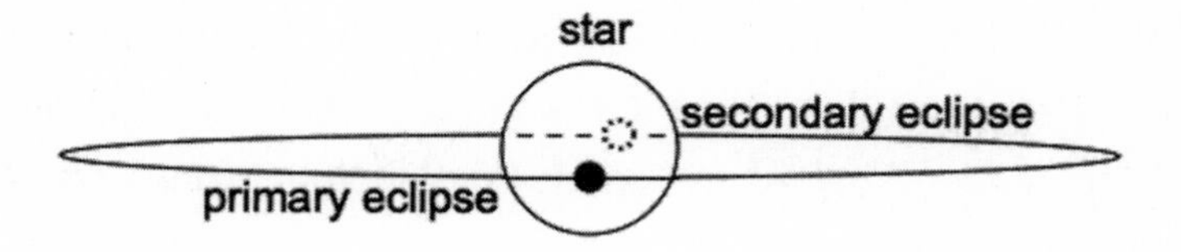

FIGURE 23. During a primary eclipse, the planet transits in front of the star, dimming the star's light. During a secondary eclipse, the planet passes behind the star, and the planet's light is blocked by the star.

telescopes would be able to take an image that showed the planet as a separate object next to the star. Spitzer was much too small to be able to accomplish that trick. All Spitzer could do was reveal that some of the light from the hot Jupiters had disappeared during the eclipse; when the planet was out of the eclipse, Spitzer had no way of telling which light had come from the planet and which from the star. Regardless, Spitzer had pushed us into a new age of direct detection of light from exoplanets, a portent of more to come when the TPFs would analyze the spectra of exoEarths in the search for biomarkers.

April 12, 2005—Michael Griffin testified at his Senate confirmation hearing that he would reassess Sean O'Keefe's decision not to send a Shuttle mission to repair Hubble. Senator Mikulski was ecstatic, and Griffin was confirmed as NASA administrator in record time that day.

May 11, 2005—Griffin sent Congress a revised budget plan for the current fiscal year (FY) that deferred the launch of both SIM and the TPFs to indefinite dates in the future and delayed the launch of

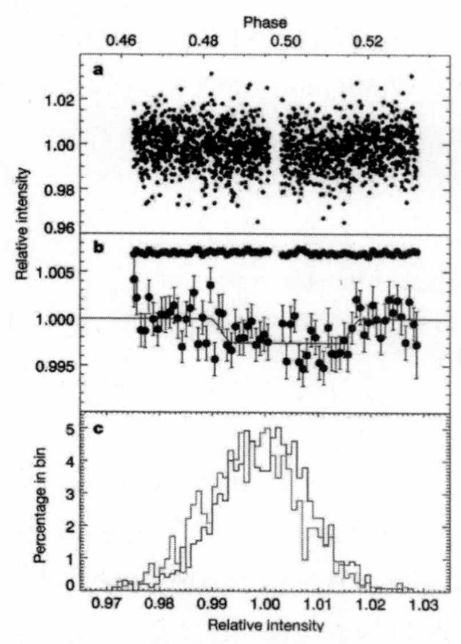

FIGURE 24. **First detection of light from an extrasolar planet, showing the dimming of HD 209458 as the planet passes behind the star during a secondary eclipse, as observed by the Spitzer Space Telescope. [Reprinted, by permission, from D. Deming et al., 2005, *Nature*, volume 434, page 741. Copyright 2005 by Macmillan Magazines Limited.]**

the Mars Science Laboratory, intended to search for microbial life on Mars, from 2009 to 2011. These deferrals and delays were caused by Griffin's need to find the money to pay for the SM4 Hubble repair mission and for cost overruns on the James Webb Space Telescope. Webb's costs had risen by $1 billion, bringing its total cost to at least $3.5 billion. In perhaps the understatement of the decade, Griffin said, "NASA cannot afford everything on its plate" and maintained that he preferred to drop lower-priority items rather than spread the pain of an inadequate budget.

President Bush had given NASA its marching orders, but he did not feel the need to support his orders for NASA with the necessary funding. Bush's Vision for Space Exploration was beginning to look like an "unfunded mandate" from the federal government to one of its own agencies.

May 24, 2005—Eugenio Rivera of UC Santa Cruz and the California-Carnegie Doppler search team submitted a paper to the *Astrophysical Journal* claiming the detection of a third planet in the Gliese 876 system. Rivera is a theorist who had worked on the Doppler data while a postdoctoral fellow at the Department of Terrestrial Magnetism. He had discovered that the data held the signature of a third planet with a minimum mass of 5.9 Earth masses, a new record for the lowest-mass planet orbiting a normal star, although Gliese 876 is an M dwarf with a mass one-third that of the Sun.

Fritz Benedict and his Hubble team had been able to detect the astrometric wobble of the first of Gliese 876's Doppler planets, Gliese 876 b, in 2002. Unlike Doppler spectroscopy, astrometry is able to determine the orientation of the orbit of a planet. The stellar wobble produced by Gliese 876 b appeared to be consistent with Gliese 876 b's orbit being within about 6 degrees of an edge-on orbit. Assuming that all three planets shared a common orbital plane, as is the case for the major planets in the Solar System, this meant that the minimum mass of Gliese 876 d was close to its true mass; the best estimate was a mass 7.5 times that of Earth. With an orbital period of 1.9 days, Gliese 876 d was the fourth hot super-Earth.

What were the hot super-Earths? Were they hot Jupiters that had lost so much hydrogen gas from being overheated by their stars that they had dropped in mass from the range of gas giants to the range of ice giants? Were they ice giants like Uranus and Neptune that had formed much farther out and then wandered inward toward their stars? Or could they be the tip of the iceberg of a heretofore unseen population of Earth-like planets? There was a strong hint about the true nature of the hot super-Earths from their sibling planets. With the single exception of the first hot super-Earth, Gliese 436 b, the hot super-Earths were known to have gas giant siblings orbiting farther out. Gliese 876 had two such gas giants, and 55 Cancri and Mu Ara each had three, with minimum masses ranging from Saturn's mass to four times the mass of Jupiter. The outermost gas giants had orbital distances comparable to Jupiter's 5.2 times the Earth–Sun distance.

This meant that the planetary systems in which three of the four hot super-Earths found themselves looked a lot like our Solar System: inner terrestrial planets surrounded by outer gas giants.

If the super-Earths had formed as ice giants in the outer regions of their planetary systems, they would have had to find some way to migrate inward past their gas giant siblings if they were to become hot super-Earths. Even though such inward migration had been suspected as the explanation for the hot Jupiters ever since the discovery of 51 Pegasi b in 1995, the fact that the hot super-Earths had outer gas giant planets argued otherwise. The Solar System appears to be a planetary system where little orbital migration has taken place, with the terrestrial planets forming inside the orbits of the gas giants, which themselves form inside the orbits of the ice giants. We can only expect the same basic outcome in other planetary systems, although the outcome may then be reworked by orbital migration. If planets migrate inward in an orderly fashion, keeping their places in the queue, then the hot Jupiters would be expected to have ice giants on longer-period, more distant orbits, and their terrestrial planets are likely to have been swallowed whole by their stars or else to have been tossed out of the way altogether. If the hot super-Earths had formed as terrestrial planets and migrated inward, they would be expected to have gas giants on longer-period orbits, as is the case with the hot super-Earths around Gliese 876, 55 Cancri, and Mu Ara.

Evidently the hot super-Earths formed inside their gas giants, just like terrestrial planets, even though their masses seemed to be out of the ballpark for Earths. In 1996 George Wetherill had published Monte Carlo models of the formation of Earth that implied that the maximum mass of a terrestrial planet was about 3 Earth-masses. However, Wetherill had assumed that the planet-forming disk was rather low in mass, an assumption that was customary at the time.

Wetherill's later work with Satoshi Inaba suggested that a disk mass a factor of 7 times higher might be necessary to form Jupiter in 4 million years. Because the final masses of terrestrial planets scale with the amount of mass available to form rocky planets, this meant that if Wetherill had run his 1996 models with an initial disk 7 times more massive, he would have found rocky planets forming with masses as high as 21 Earth-masses—high enough to explain the probable true masses of the four hot super-Earths.

Doppler spectroscopy, the leader of the pack since 1995, appeared to be in the process of detecting not just Jupiters, but now Earths, however overweight and overheated they might be.

September 2, 2005—A panel led by Space Telescope Science Institute astronomers charged with figuring out how to cut down on the burgeoning costs of the Webb Telescope insisted that neither reducing the 260-inch (6.5-meter) mirror size nor eliminating a major instrument was an acceptable option. The panel did agree that some testing and polishing could be eliminated, the launch could be delayed from 2011 to 2013, and a French-built coronagraph could be dropped altogether. The coronagraph would have allowed Webb to at least attempt to image exoplanets, even if it had no real hope of seeing Earths.

September 28, 2005—Jean-Phillippe Beaulieu of the Institute of Astrophysics in Paris, David Bennett, and a cast of thousands (71 coauthors) submitted a paper to *Nature* announcing a new microlensing planet discovery. This time the planet's mass was estimated to be only 5.5 Earth masses, the lowest estimate for any planet outside the pulsar planet system.

The OGLE 2005-BLG-390L microlensing event in July and August 2005 had yielded a small but distinct blip in the brightness of the background star lasting less than 24 hours. Thanks to the around-the-clock coverage provided by the worldwide resources of OGLE, MOA, and two French groups (PLANET and RoboNet), the microlensing network had obtained a good sampling of the entire short-lived planetary blip, ruling out other interpretations of what caused the brief brightening.

Assuming that OGLE 2005-BLG-390L b orbited an M dwarf star with about one-fifth the mass of the Sun, then it orbited its star at a distance of 2.6 AU, equivalent to the middle of the asteroid belt in our Solar System. Considering the faintness of such an M dwarf, OGLE 2005-BLG-390L b was probably a chilly world, the first "cold super-Earth," with a surface temperature similar to that of Neptune or Pluto. OGLE 2005-BLG-390L b was not a world you would want to visit without wearing some serious thermal underwear.

OGLE 2005-BLG-390L b was the third planet detection by microlensing. Andrzej Udalski and a mere 32 co-authors had detected the second microlensing planet, OGLE 2005-BLG-071 b, in March 2005. It was the second gas giant with a mass estimated at 2.7 Jupiter masses orbiting about 3 AU from an M dwarf star. Years of searching had yielded two gas giants and one cold super-Earth. Considering that the blips caused by Jupiter-mass planets lasted for several days to a week, whereas the 5.5-Earth-mass planet bent the background star's light for less than a day, bigger planets were easier to detect than smaller planets. The microlensing folks expected to find a lot more Jupiters than Earths, if they were equally common inhabitants at the distance from an M dwarf where they were most likely to be detectable—namely, in the "Einstein ring" lying at asteroidal distances from M dwarfs.

The fact that microlensing's score card included only two gas giants and one cold super-Earth implied that cold super-Earths were more frequent than gas giants, or else the microlensing teams had been incredibly lucky to have found the 5.5-Earth-mass planet. Beaulieu and his colleagues did not feel particularly lucky. So they decided that their detection had shown that cold super-Earths were much more common around M dwarfs than were gas giants, and they interpreted this as proof that core accretion was the dominant formation mechanism for gas giant planets.

Greg Laughlin and Peter Bodenheimer of UC Santa Cruz, along with Fred Adams of the University of Michigan, had published a paper in the *Astrophysical Journal* in 2004 showing that core accretion was unlikely to form gas giants around M dwarf stars, because core accretion proceeded even more slowly around such stars than around solar-type, G dwarf stars. Orbital periods are longer around M dwarfs than around solar-type stars at the same distance. The Dodge-Em cars did not run as fast in the M dwarf ride as they did in the G dwarf ride, making it harder to effect a decent rear-end collision. Laughlin and his colleagues decided that core accretion was more likely to result in failed cores such as Neptune around M dwarfs, rather than in Jupiters.

The microlensers evidently were not aware of a second paper on core accretion submitted to *Astronomy & Astrophysics* on June 23, 2005, by Kacper Kornet of the Max Planck Institute for Astronomy in Heidelberg and his colleagues. This paper made the opposite prediction. Using a somewhat different approach and assumptions, Kornet had shown that core accretion should lead to gas giant planets being even more frequent around lower-mass stars than around higher-mass stars. If this were true, then core accretion was effectively ruled out by the microlensing detections. Which was it?

Regardless of the implications for the ongoing debate, the discovery of OGLE 2005-BLG-390L b meant that we now had evidence for extrasolar analogues of all three major classes of planets in the Solar System. The hot super-Earths, hot and warm Jupiters, and cold super-Earths appeared to be alien versions of the terrestrial planets, gas giants, and ice giants, respectively, in our Solar System. All of this was achieved within the 10 years since Mayor and Queloz had discovered 51 Pegasi b; we had gone from nothing to everything in a decade.

October 21, 2005—*Science* reported that Michael Griffin was "fed up" with receiving conflicting advice from scientists about how to solve the problems afflicting NASA's science budget. He was particularly upset about scientists lobbying Congress in an effort to save their projects, such as Hubble. Griffin did promise not to divert any NASA science funding to solve the budget problems involving human space flight at NASA.

December 9, 2005—NASA headquarters further delayed the launch of the Webb Telescope from 2014 to 2016. The extra costs involved in adding two more years of work on Webb used up all the cost savings gained as a result of the Webb descope exercise a few months earlier. The total cost of Webb was thus still $4.5 billion. Given the zero-sum game in NASA's Science Mission Directorate, it was clear that the funds needed for Webb would have to come from other NASA science missions. Although Webb could not see exoEarths, it would be able to see newly formed Jupiters orbiting nearby young stars, if any existed.

February 6, 2006—Fritz Benedict submitted a paper to the *Astrophysical Journal* that combined Doppler data and astrometric measurements, taken by the Hubble Space Telescope and by George Gatewood's Allegheny Observatory telescope, of the nearby star Epsilon Eridani. This star had a long history of claims for planet detections dating back to Peter van de Kamp in 1974. Bruce Campbell and Gordon Walker had first thought that Epsilon Eridani might have a giant planet, but they backed off this claim in their 1988 paper. Benedict's new data implied that Epsilon Eridani had a planet with a mass of 1.6 Jupiter masses, orbiting about 2 AU from its star every 6.9 years. It was similar to the planet inferred almost two decades earlier by Campbell and Walker and their team. As in the case of Gamma Cephei's planet, it looked like the Canadians had been there first but could not be certain what they had discovered.

March 2, 2006—Dennis Overbye of the *New York Times* reported that NASA administrator Michael Griffin had raided the budget of NASA's Science Mission Directorate (SMD) to the tune of $3 billion over the next 5 years in order to help pay for the highest-priority items in President Bush's Vision for NASA: returning the Space Shuttle to flight, finishing construction of the International Space Station, and developing the new Crew Exploration Vehicle and the associated rockets needed for the return to the Moon and the trip to Mars. This diversion of science funds to human space flight was perpetrated in spite of Griffin's earlier promises to Congress that not "one thin dime" would be taken from NASA's space science programs to solve NASA's looming budget crisis. One dime may be thin, but $3 billion worth of dimes, stacked together, would stretch a distance of about 30,000 miles, more than enough to wrap around the entire Earth at its equator.

Griffin had replaced veteran bureaucrat Al Diaz with Mary Cleave as the head of NASA's SMD. Whereas Al Diaz received his bachelor's and master's degrees in physics, and his predecessor Ed Weiler holds a Ph.D. in astrophysics, Cleave's credentials included a Ph.D. in civil and environmental engineering and field experience with the flow of water in desert basins—not exactly what one would expect for the head of NASA's science efforts. However, Cleave is also a former astronaut who was rumored to have rallied the astronaut corps behind Griffin's candidacy for NASA administrator. Mary Cleave's summary of the loss of $3 billion from her SMD budget was simple and elegant: "We took a hit."

Indeed "we" did. As a direct result of the $3 billion cut, the Terrestrial Planet Finder missions were "deferred indefinitely," as was the Mars Sample Return mission. The estimated cost for the TPF program had risen from $4.5 to $6.5 billion, and now that it was deferred indefinitely, the funding dropped precipitously to zero. TPF project scientist Chas Beichman said, "We're getting ready to fire all the people we have built up," many of them young scientists eager to begin a career in planet hunting. The estimated cost for SIM had by now risen to $4 billion, into the realm of flagship NASA missions and just shy of the estimated cost of $4.5 billion for the Webb Telescope. As a result of the $3 billion cut, SIM's funding was decreased by half and the mission was delayed 3 more years to a 2015 or 2016 launch.

Cancellations and delays spread the intense budgetary pain throughout NASA's science portfolio, even down to the level of research and analysis programs, the grants awarded to individual scientists. The competition for a grant from a NASA Research & Analysis (R&A) program is fierce, and typical success rates are on the order of 10% for new starts. Having been awarded such a 3-year grant, roughly a year after submitting a detailed proposal, the recipient has 2 years to get the work done before it is time to submit a new proposal for the next 3 years of work. The perpetual shortage of research funds has often meant that the budgets for NASA's R&A programs were held constant, without inflation adjustments, or hit with one-time "taxes" of 5% to solve problems elsewhere in the science portfolio. But the $3 billion cut led to the most drastic measures ever experienced: 15% cuts across the board. The R&A program that had been intended to build the scientific foundation for TPF was canceled outright, and the funds for the fledgling NASA astrobiology program were cut in half. It looked like it was time to abandon ship.

March 10, 2006—Ohio State University's Andrew Gould submitted a paper to the *Astrophysical Journal* Letters reporting the fourth detection of an exoplanet by the gravitational microlensing method. Amazingly, Gould and his 35 colleagues had discovered a second super-Earth during the OGLE 2005-BLG-169 microlensing event the year before. The planet's mass was roughly 13 Earth masses, if its host star had a mass of about half a solar mass, as was expected for an M dwarf host star. The OGLE 2005-BLG-169 b planet was orbiting at an asteroid-like distance of about 3 AU, making it another cold super-Earth.

The fact that half of the four microlensing events had netted cold super-Earths, while half had yielded cold Jupiters, seemed to bear out the suspicion Jean-Phillippe Beaulieu had expressed the previous year that cold super-Earths were much more common at asteroidal distances around low-mass M dwarfs than were gas giant planets. Once again, the jury decided to return a verdict of "guilty" for core accretion as the dominant formation mechanism for Jupiters. Disk instability was not even listed as an accomplice in the indictment handed down by the 36-scientist microlensing jury.

The two microlensing discoveries of failed cores seemed to prove that core accretion was at work, at least around M dwarfs, the most common stars. I learned about Gould's discovery while attending a conference in Houston, Texas, and was somewhat chagrined by the negative implications for disk instability in this most recent trial.

The implications of the cold super-Earths were rumbling around my brain as I surfed the Internet in the lobby of the Hilton Hotel in Clear Lake, Texas, across the road from NASA's Johnson Space Center. I had another "ah-ha!" moment, as I had while writing an e-mail to the *Washington Post*'s science reporter in 2001. It occurred to me that the cold super-Earths could be explained just as well by disk instability as by core accretion.

In 2001 I had imagined that Uranus and Neptune had started out as giant gaseous protoplanets, formed by disk instability, which had been whittled down to ice giants by the loss of most of their hydrogen and helium gas envelopes. The gaseous envelopes would be lost once the ultraviolet light from nearby, newly formed massive stars had stripped away the outer disk gas and then begun to denude the outermost protoplanets. The key point was that this process occurs only outside a critical radius that depends on the mass of the central star. Inside this radius, the star's gravity is too strong for the hot gas to be stripped away. The critical radius depends directly on the mass of the star, so for the M dwarf stars being probed by microlensing, the critical radius was expected to be about 2 to 10 times smaller than that for our Sun. Given that the critical radius was expected to be roughly at Saturn's distance in the solar nebula, this meant that for M dwarfs the critical radius should be roughly at asteroidal distances: 2 or 3 AU from the M dwarfs. Checkmate. That was the estimated orbital distance of the cold super-Earths. These ice giants could have started out as protoplanets formed by disk instability, only to be deconstructed into cold super-Earths.

The fact that the microlensing surveys had found strong evidence for many more cold super-Earths than cold Jupiters at asteroidal distances around M dwarfs was consistent with the fact that most stars form in regions of high-mass (O, B) star formation, where the outer disks are subject to the ultraviolet strip-search process. The much smaller fraction of M dwarfs that form away from such regions would be expected to have their outer gas giant protoplanets survive to become gas giant planets—that is, the cold Jupiters seen by microlensing. It was a "Just So Story" worthy of Rudyard Kipling himself, though, I fervently hoped, not quite so fanciful.

I returned to Washington, D.C., and immediately wrote a paper on cold super-Earths and submitted it to the *Astrophysical Journal* Letters on March 31. If correct, this novel explanation further supported the hope that planetary system formation could proceed in regions where massive stars form. Such regions need not be inimical to the planet formation process, it spite of earlier fears that planets could not form in such hostile environments. Instead, this heretical scenario supported the idea that planetary systems, even those similar to our own, can form in just about any star-forming region, making the total number of planetary systems in the Galaxy immensely higher than would be the case if special circumstances were required for planets to form. Earths would be nearly everywhere in this theoretically crowded universe. The challenge was to find them, and for that we needed NASA's dream of dedicated planet-hunting space telescopes.

July 7, 2006—In order to solve the latest short-term funding problem for NASA's Science Mission Directorate, Griffin announced that he had decided to "refocus" SIM and further delay its launch to well beyond the hoped-for 2015 or 2016. In 2003, SIM's planned launch date was only 6 years off, at the end of 2009. Three years later, SIM's launch date had slipped by at least another 7 years, more than the traditional slip of about 1 year per year. SIM seemed to be receding from Earth, with a large and accelerating Doppler redshift, an indication that it soon might never be seen again.

July 19, 2006—George Wetherill died, without knowing for certain how Jupiter formed, how many Earth-like worlds exist, or how common life is in the universe.

CHAPTER 6

The Battle of Prague

Basic research is like shooting an arrow into the air and, where it lands, painting a target.

—HOMER ADKINS (1984)

August 24, 2006—In an extraordinary final plenary session at the International Astronomical Union's triennial General Assembly in Prague, Czech Republic, the IAU's astronomers voted to pull Pluto from the nine-member team of the Solar System's planets. The 400-odd astronomers present had voted overwhelmingly in favor of the move by waving yellow cards that signified IAU members, making the plenary hall look as if it were full of demented soccer referees cautioning an overly aggressive player. Given that they were in fact ejecting Pluto from the game, they should have been waving red cards.

Pluto was discovered on February 18, 1930, by Clyde Tombaugh at the Lowell Observatory in Flagstaff, Arizona. Tombaugh had been searching for the hypothesized "Planet X" that was thought to be disturbing the motions of the outermost planets Uranus and Neptune. He had been looking for only a year when he stumbled across

Pluto. The founder of the Lowell Observatory, Percival Lowell, had hypothesized the existence of Planet X and searched for it himself, but he had died in 1916. Fourteen years later, Tombaugh struck pay dirt in the arid northern Arizona hills.

Tombaugh had found Pluto by the same method that the Terrestrial Planet Finders intended to use to discover new Earths: direct imaging. In the case of Pluto, the task was immensely simpler than it is for extrasolar Earths, not only because Pluto is much brighter than a distant Earth but also because Pluto's light need not be blocked by the light from the Sun, at least not during the portions of Earth's orbit when Earth is in between the Sun and Pluto. To look for Pluto, then, Tombaugh simply pointed Lowell's new 13-inch telescope in the opposite direction from the Sun, close to the same orbital plane where the other eight planets orbit, and started taking pictures on glass photographic plates. By comparing two plates that he had taken six days apart, Tombaugh could look for anything that jumped to a new location on the plates during that time interval. The jump would be caused not by Pluto's own motion across the sky, which is painfully slow given its 248-year orbital period around the Sun. Instead, the apparent shift in Pluto's position would be caused by Earth's much more rapid orbit around the Sun. Any nearby object will appear to move with respect to distant background objects if the observer is moving. A simple way to demonstrate this is to hold out your arm and stare at your thumb as you blink first one eye and then the other. When you do so, your thumb appears to jump back and forth, just as Pluto did for Tombaugh in 1930. This parallax effect made it possible to determine the distance of Pluto: its mean distance from the Sun of 39 AU is 39 times greater than that of Earth. Pluto was by far the most distant planet in the Solar System, easily beating Neptune's average distance of about 30 AU.

Tombaugh continued his search for new planets in the outer Solar System, but 13 more years of searching yielded nothing. Nevertheless, Pluto was the ninth planet for 76 years. Until the pulsar planets were found in 1992, Tombaugh was the only living person to have found a new planet anywhere in the universe. The exclusive club of planet finders became increasingly crowded after 51 Pegasi b's discovery in 1995, and by 2006 its membership had swelled to hundreds of astronomers worldwide.

Why was Pluto ceremoniously demoted from planethood? The most immediate reason was the discovery, in mid-2005, of a new object on the edge of the Solar System dubbed 2003 UB313 by Caltech's Michael Brown and his colleagues. Using the 48-inch (1.2-meter) Oschin telescope at the Palomar Observatory in southern California, Brown had employed the same basic approach as Tombaugh, except for the replacement of photographic plates by the electronic devices that make digital cameras possible. The fact that Brown could store his images on a computer meant that the computer could also be trained to "blink" the images taken on different nights, searching for anything that jumped around. Tombaugh had been forced to do his own blinking by eye, as well as to do his own observing, but Brown had the luxury of a fully automated, robotic telescope and data analysis system, which meant that any newly discovered objects were waiting for him when he arrived at his Caltech office each morning.

2003 UB313 is currently the most distant known member of the Solar System, with a mean distance from the Sun of about 65 AU and an orbital period of 560 years, over twice that of Pluto. Astoundingly, this distant object is relatively bright, which means that it must be fairly large. When Brown and his team announced their discovery in July 2005, they estimated that it might even be larger than Pluto.

FIGURE 25. Pluto and its satellites Charon, S/2005 P1, and S/2005 P2, as observed by the Hubble Space Telescope. [Courtesy of NASA, ESA, Harold Weaver (Johns Hopkins University, Applied Physics Laboratory), Alan Stern (SwRI), and the HST Pluto companion search team.]

Subsequent detailed measurements by Hubble proved their initial estimate to be correct.

But if 2003 UB313 was larger than Pluto, was it a planet too? If so, it would be up to the IAU to decide, because the IAU has the ultimate responsibility for categorizing and naming celestial bodies. If the IAU agreed, Michael Brown, Chad Trujillo, and David Rabinowitz had just found the tenth planet. Whereas Tombaugh's planet had been named Pluto, after the mythological god of the underworld, Brown had playfully given 2003 UB313 the name Xena, after the warrior princess on a popular television series. Soon after the initial discovery, astronomers realized that Xena had a satellite, and Brown named the satellite Gabrielle, for Xena's television sidekick.

The IAU had been wrestling with the issue of planethood well before Xena made her appearance on stage. A Working Group on the Definition of a Planet had been created in June 2004 to deal with the problem of Pluto's planethood. It had become clear that Pluto was the largest known member of the swarm of comets in the Kuiper Belt, a region named after the same Gerard Kuiper who had first suggested gravitational instability as a means for forming gas giant planets. Many Kuiper Belt objects have orbits similar to that of Pluto, and

RCW 49 is a cluster of several thousand newly formed stars in our Milky Way Galaxy, imaged here in infrared light by the Spitzer Space Telescope. Massive stars (blue) lie at the center of the cluster, surrounded by clouds of gas and dust (pink) that will form new stars and planets. [Courtesy of NASA, JPL-Caltech, and E. Churchwell (University of Wisconsin).]

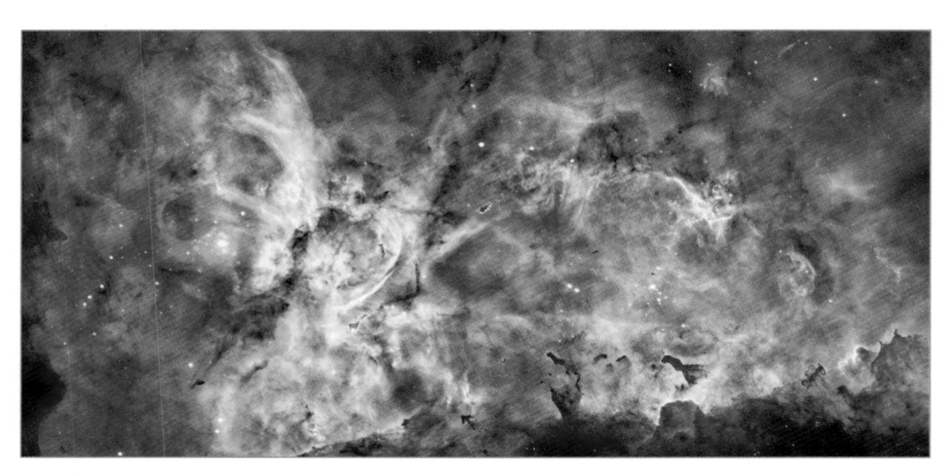

The Eta Carina nebula contains 60 massive O stars and thousands of lower mass stars, imaged here in optical light by the Hubble Space Telescope. The stars are immersed in clouds of gas and dust in the process of forming new stellar and planetary systems. [Courtesy of NASA, ESA, N. Smith (University of California, Berkeley), and the Hubble Heritage Team (STScI/AURA).]

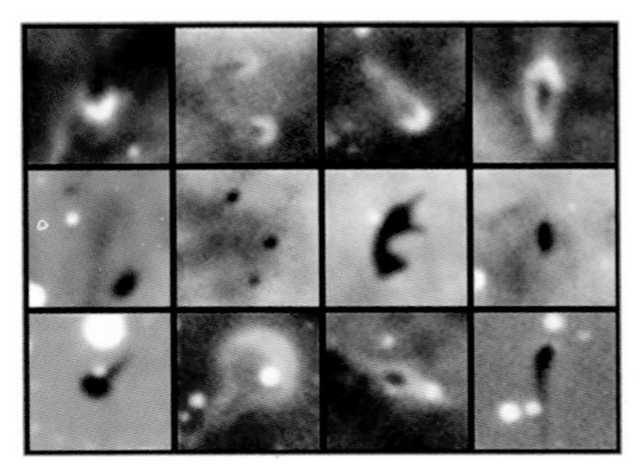

Young stars and their protoplanetary disks in the Eta Carina nebula, imaged in optical light by the 4-m telescope at the Cerro Tololo Inter-American Observatory in Chile. Harsh ultraviolet light from massive stars in Eta Carina heats the gas around these young stars and their disks to temperatures high enough to strip away the gas, giving them a comet-like appearance. [Courtesy of Nathan Smith (U. C. Berkeley) and John Bally (U. Colorado).]

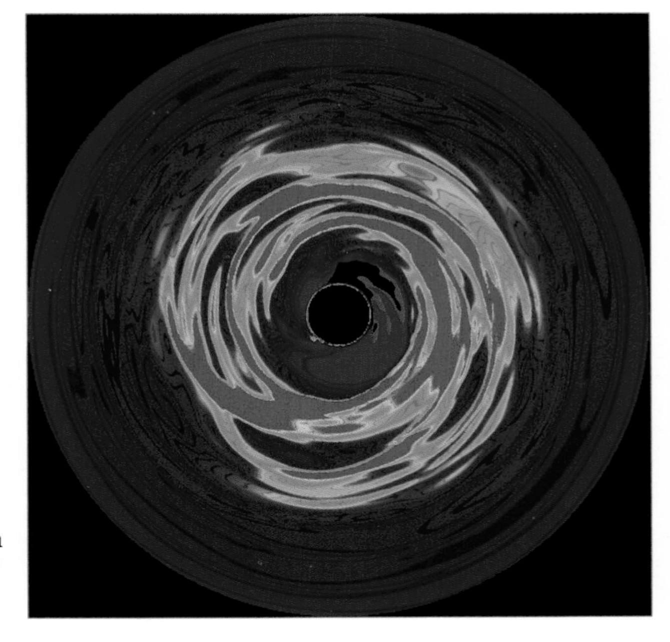

Theoretical model of the solar nebula, the protoplanetary disk that formed our Solar System. A solar-mass protostar lies unseen at the center of this false-color image of the gas and dust in the midplane of the disk. Orange corresponds to high densities, black and purple to low densities. The outer edge lies at the orbital distance of Uranus. In the disk instability mechanism of giant planet formation, the spiral arms form dense clumps that contract to form the giant planets. [Model by Alan Boss.]

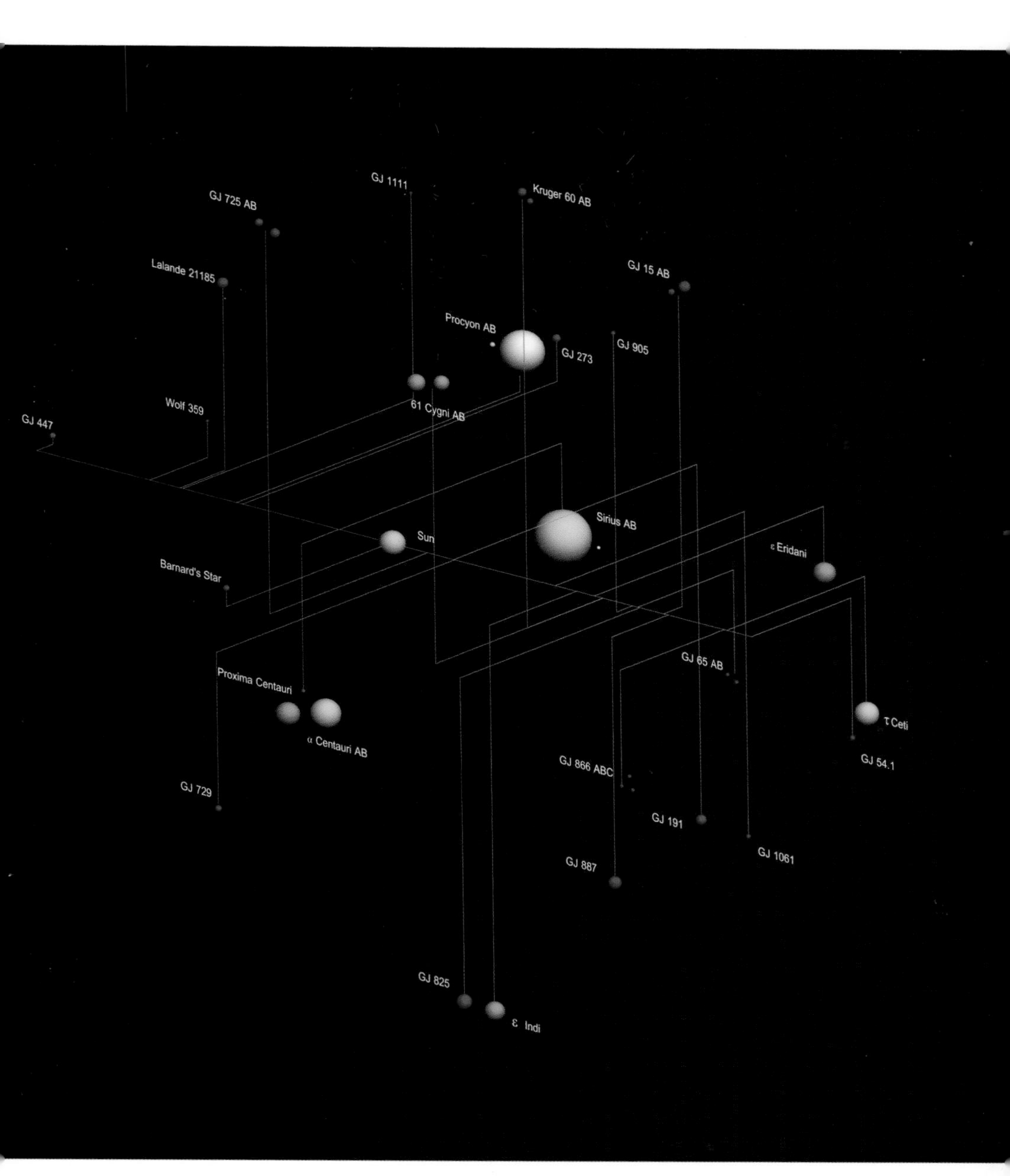

The 25 stars closest to the Sun in our neighborhood of the Milky Way Galaxy. Most of these stars are red dwarf stars (M dwarfs) with masses less than the mass of our Sun. [Courtesy of T. Henry (Georgia State University) and the RECONS Project.]

Artist's conception of the gas giant planet (with several hypothetical moons) orbiting the M dwarf star Gliese 876. [Courtesy of NASA and G. Bacon (STScI).]

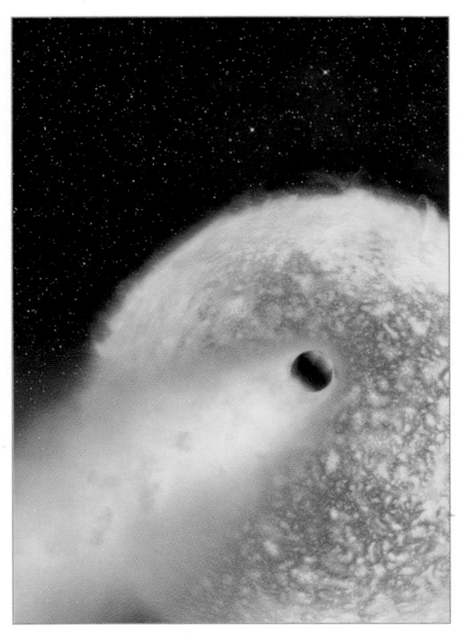

Artist's conception of hot gases flowing off the atmosphere of the first transiting planet, the hot Jupiter orbiting the star HD 209458. [Courtesy of ESA, Alfred Vidal-Madjar (Institute of Astrophysics, France), and NASA.]

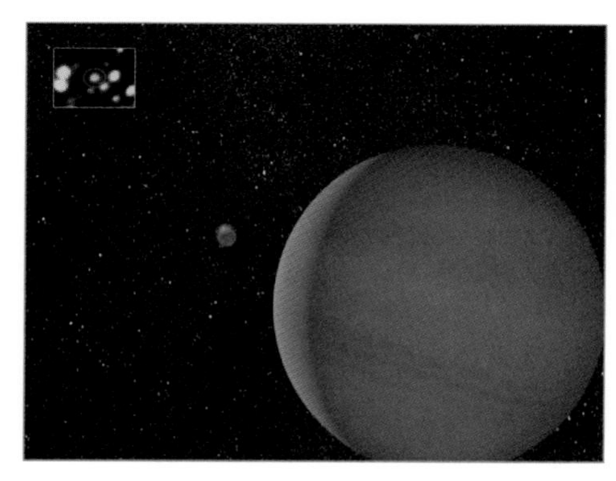

Artist's conception of the planet orbiting an M dwarf star that produced the first detection of an extrasolar planet by the gravitational microlensing technique, OGLE 2003-BLG-235/MOA 2003-BLG-53 b. The inset shows an image of the background star that brightened during the microlensing event. [Courtesy of Las Campanas Observatory (Carnegie Institution), NASA, and JPL-Caltech.]

Artist's conception of the hot super-Earth found by the Doppler wobble technique around the solar-type star 55 Cancri. [Courtesy of NASA.]

Artist's conception of the first cold super-Earth, a planet with a mass of about 5.5 Earth masses, detected in the microlensing event OGLE 2005-BLG-390L, orbiting a red dwarf star. [Courtesy of NASA, ESA, and G. Bacon (STScI).]

Artist's conception of the first hot super-Earth discovered by the Doppler wobble technique, orbiting the M dwarf Gliese 436. This planet is also the first hot super-Earth found to be a transiting planet, allowing its density to be measured. [Courtesy of NASA.]

Artist's conception of the hot Jupiters orbiting the stars HD 209458 and HD 189733. The light emitted by these planets at a range of infrared wavelengths was first measured by the Spitzer Space Telescope. [Courtesy of NASA, JPL-Caltech, and T. Pyle (SSC).]

NASA's Hubble Space Telescope, launched into low-Earth orbit in 1990, is seen against the Earth's clouds and oceans. [Courtesy of NASA and STScI.]

Artist's conception of NASA's Spitzer Space Telescope (SST), launched into an Earth-trailing orbit in 2003. SST is the largest and most sensitive infrared astronomical space telescope ever launched. [Courtesy of NASA, SSC, and JPL-Caltech.]

Artist's conception of the Canadian MOST space telescope, launched into a low-Earth orbit in 2003, which has been used to study transiting extrasolar planets. [Courtesy of Jaymie Matthews (U. British Columbia) and MOST.]

Artist's conception of the French space telescope COROT in low-Earth orbit. COROT was designed to discover extrasolar planets as small as super-Earths by the transit detection technique. [Used with permission of by CNES.]

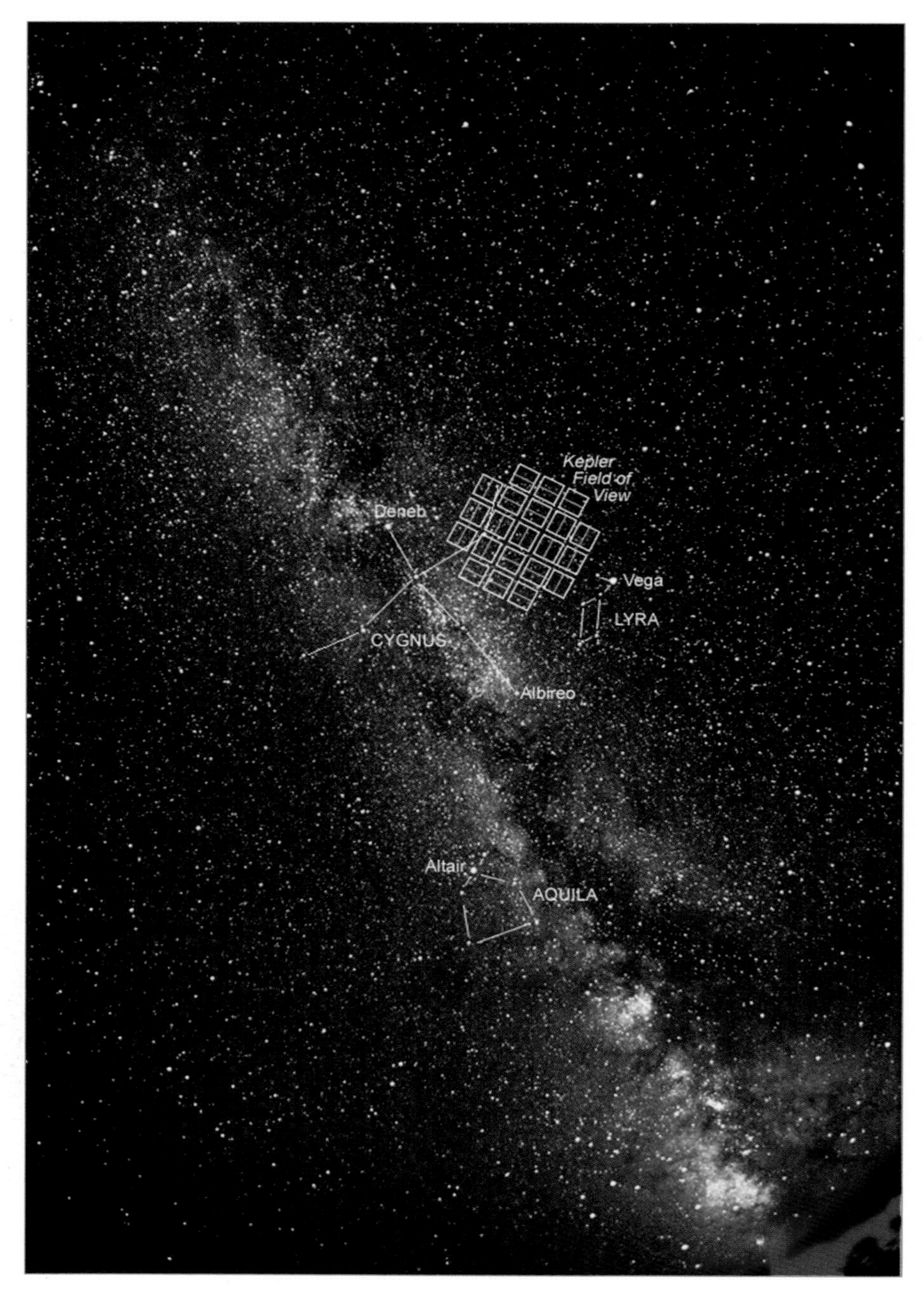

NASA's Kepler Mission will search for Earth-like worlds around at least 100,000 stars in the constellation Cygnus. This image shows the fields of view to be searched by Kepler, located above the midplane of our Milky Way Galaxy. [Courtesy of the Kepler Mission, NASA Ames Research Center.]

Artist's conception of the Kepler Space Telescope, to be launched by NASA into an Earth-trailing orbit in early 2009. Kepler was designed to detect Earth-like, habitable worlds by the transit detection technique. [Courtesy of the Kepler Mission, NASA Ames Research Center.]

NASA's James Webb Space Telescope (JWST), currently planned for launch in 2013, will be the largest (6.5-m) infrared space telescope ever built. This artist's conception shows JWST along with a protoplanetary system in the process of formation. [Courtesy of NASA and STScI.]

Artist's conception of the Space Interferometry Mission, a NASA space telescope that would be able to detect Earth-mass planets in the habitable zone of nearby stars. SIM would detect the astrometric wobble of the star induced by the presence of habitable planets. [Courtesy of NASA and JPL-Caltech.]

Artist's conception of the Terrestrial Planet Finder Coronagraph, a NASA space telescope that would be able to image and study Earth-like planets around nearby stars. TPF-C would detect habitable planets at optical wavelengths and search their reflected starlight for biomarkers indicative of the presence of life. [Courtesy of NASA and JPL-Caltech.]

Artist's conception of the Terrestrial Planet Finder Interferometer, a NASA space telescope that would be able to image and study Earth-like planets around nearby stars. TPF-I would detect habitable planets at infrared wavelengths and search their emitted light for biomarkers indicative of the presence of life. [Courtesy of NASA and JPL-Caltech.]

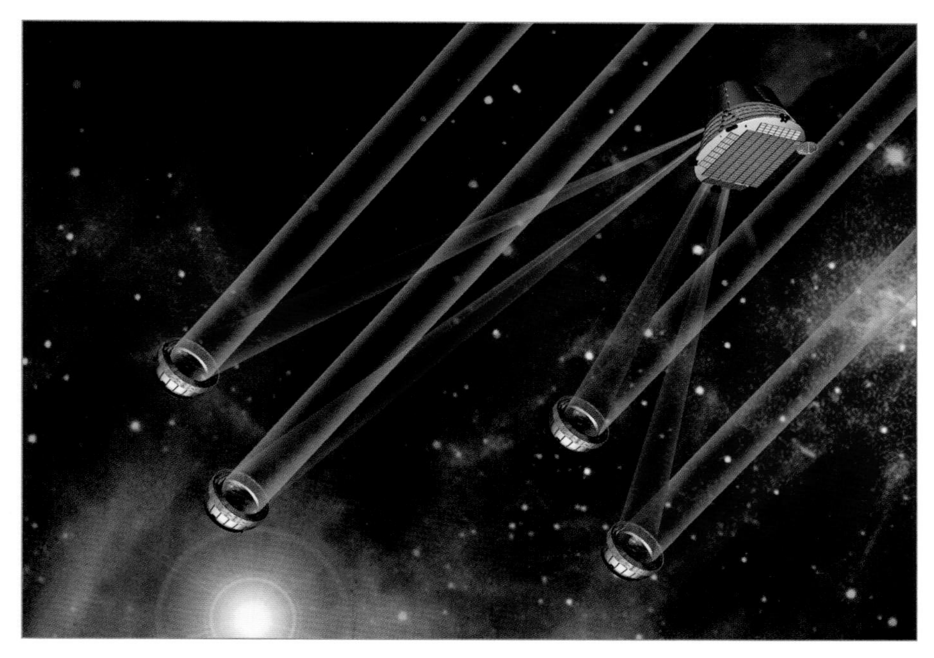

Artist's conception of the Emma X-array version of the joint NASA-ESA TPF-I/Darwin Mission to detect and study Earth-like planets. This version is named after Emma Darwin, wife of Charles Darwin, and consists of four collector spacecraft and a fifth combiner spacecraft, located out of the plane of the four collector spacecraft. [Courtesy of NASA and JPL-Caltech.]

Michel Mayor, the leader of the Swiss Geneva Observatory planet-hunting team. [Courtesy of Michel Mayor.]

Paul Butler, the leader of the Carnegie planet-finding team. [Courtesy of R. P. Butler.]

Geoffrey Marcy, the leader of the California planet-finding team. [Courtesy of G. Marcy.]

Tim Brown, a leader in transit detections of extrasolar planets, is shown with the STARE telescope [Courtesy of UCAR and Carlye Calvin.]

William Borucki, the originator and science team leader of NASA's Kepler Mission to determine the frequency of Earth-like planets in the Milky Way Galaxy. [Courtesy of NASA Ames Research Center.]

Malcolm Fridlund, project scientist for the European COROT and Darwin Missions to detect and characterize habitable planets. [Courtesy of M. Fridlund. Copyright 2006 by Skolastro.]

Michael Shao, the originator and Project Scientist for NASA's Space Interferometry Mission, which would detect Earth-like planets around nearby stars. [Courtesy of NASA and JPL-Caltech.]

David Charbonneau, Drake Deming, Kim Weaver, and Alan Boss (left to right) during the 2005 NASA HQ press conference announcing the first detection of light from an extrasolar planet. [Courtesy of NASA.]

Artist's conception of the MOA-2007-BLG-192L system, where microlensing revealed the presence of a 3-Earth-mass planet in orbit around a brown dwarf star. [Courtesy of NASA's Exoplanet Exploration Program.}

Artist's conception of the trio of super-Earths found by Doppler spectroscopy to orbit the star HD 40307. [Courtesy of ESO.]

they had been found by the hundreds, making the situation similar to that of Ceres in the asteroid belt. Ceres was discovered in 1801 by the Sicilian astronomer Giuseppe Piazzi. At first it appeared that Ceres was the long-sought "missing planet" between Mars and Jupiter, but then another object with a similar orbit, Pallas, was found in 1802; a third, Juno, in 1804; and a fourth, Vesta, in 1807. By this time it was clear that Ceres was just one of a number of objects orbiting between Mars and Jupiter, and these became known as the minor planets in recognition of their diminutive stature compared to the eight major planets. We now know that the asteroid belt is populated by many hundreds of thousands of bodies large enough to be detected from Earth.

The same fate had befallen Pluto. It was merely the first—and so far the largest, until Xena—body detected in the Kuiper Belt. By the same reasoning that had led to the demise of Ceres's claims to planethood in the early 1800s, it seemed that Pluto deserved a similar demotion.

The IAU Working Group on Extrasolar Planets had agreed in 2001 on a "working definition" of a planet that could be applied to extrasolar planetary systems. However, we were focused on the upper-mass end: How massive could an object in orbit around a star be and still be called a planet? What if a Jupiter-mass object were not in orbit around a star? Would it still be a planet? Our main debate had centered on whether planets should be defined by their formation mechanism or by their ability to undergo nuclear fusion.

Stars form from the gravitational collapse of clouds of gas and dust that swirl around the Galaxy, after having been ejected from previous generations of stars by their stellar winds or by supernova explosions. Given a small initial amount of rotation, a collapsing cloud will flatten into a disk, with a growing protostar at its center. If the rotation rate is high enough, the cloud will collapse and fragment into two or more protostars, each surrounded by a rotating disk of gas and dust. Planets form from the leftovers of the star formation process, in the portions of the gas and dust disk that manage to escape being swallowed by the omnivorous, voracious central protostar. Although planet formation is thus intimately linked to star formation, the physics of planet formation is intrinsically different from that of star formation, even in the case of gas giant planet formation by disk instability. Binary and multiple protostars can form from gravitational instabilities in a newly formed, massive protostellar disk, but disk instability entails the formation of gaseous protoplanets in a protoplanetary disk that contains only a small fraction of the mass of the already formed, central protostar. Planets therefore begin their formation only well after their stars have largely formed.

Stars are defined as objects large enough to sustain thermonuclear reactions deep in their interiors, such as the fusion of hydrogen nuclei (protons) into helium nuclei. Hydrogen fusion is the source of the energy that empowers stars like the Sun to shine brightly for billions of years. Stars with masses less than about 75 Jupiter masses, however, do not become hot enough in their centers for hydrogen fusion to take place. As long as they are more massive than about 13 Jupiter masses, though, they can burn deuterium, which is hydrogen with the addition of a neutron to its nucleus. Deuterium is a relatively rare isotope, so deuterium burning is able to sustain fusion reactions only for millions of years, instead of billions of years.

The Working Group on Extrasolar Planets therefore decided that planets must be objects in orbit around stars (or stellar remnants, such as pulsars) with masses less than about 13 Jupiter masses. Objects with masses in the range from 13 to 75 Jupiter masses would be brown dwarfs, even if they managed to form by the same process (such as disk instabilty) that formed a sibling object with a mass of 12 Jupiter masses. Isolated objects not in orbit around a star but moving on their own in interstellar space with masses less than 13 Jupiters were not planets but "sub-brown dwarfs." This is because they presumably formed by the same collapse and fragmentation process that forms stars and brown dwarfs, even though they failed to reach the critical mass needed to burn deuterium. Achieving agreement on this working definition was difficult, but we accomplished the task in a few months of e-mail traffic. We had defined the upper-mass end of planethood, but we avoided the lower mass end, despite the fact that one of the pulsar planets has a mass only slightly larger than that of the Moon, about 1/67 the mass of Earth. We knew that the best context for debating the lower-mass end was the Solar System, so the Working Group left that decision for the IAU's Solar System specialists.

The IAU proceeded to delve into the fate of Pluto in March 2005, prior to the discovery of Xena but a few months after Alex Wolszczan had presented evidence for a possible fourth pulsar planet with a mass only one-fifth that of Pluto. It was time to get serious about the low-mass end of the planet spectrum.

The Working Group consisted of 15 astronomers who headed various IAU planetary science committees, as well as 4 others, including Brian Marsden, head of the Minor Planet Center at the Smithsonian Astrophysical Observatory in Cambridge, Massachusetts, and Alan Stern, director of the Southwest Research Institute's Department of Space Studies in Boulder, Colorado. Marsden's Center is charged with keeping track of the myriad minor planets, the denizens of the Kuiper Belt, and anything else that is found wandering dazed around the Solar System pushing a shopping cart. Stern was the leader of NASA's first robotic mission to Pluto, New Horizons, already launched and slated to shoot past Pluto in 2015 with its camera madly clicking photos, like an eager tourist trapped on an express bus that will not stop as it passes Mount Rushmore.

Stern had argued in a *Sky & Telescope* article in 2002 that any object that was massive enough to be round should be called a planet. Roundness is enforced by gravity, which tends to pull objects into spherical shapes. Most asteroids and comets are not round but, rather, have odd shapes that more often resemble potatoes than baseballs. Their small sizes and masses mean that gravity is not able to overcome the internal strengths of the rocks and ices of which they are made. If roundness were the criterion for planethood, then the Solar System would have a lot more than nine planets: the minor planets Ceres, Vesta, and Pallas might be promoted to full planethood, although Juno was likely to be left as a minor planet. Pluto also fit comfortably within the roundness definition. Stern was not about to propose that his New Horizons mission, billed as the first spacecraft to visit the most distant planet, was not really headed for a planet after all.

Because Marsden was presented daily with evidence that Pluto was the largest known member of a growing swarm of increasingly large Kuiper Belt comets, he might have been expected to have a somewhat different point of view than Stern on the issue of Pluto's planethood. It is often said that the outcome of a committee's work is decided by who is chosen to sit on the committee. The same could be said of the IAU's Working Group on the Definition of a Planet. The membership was divided among those who wished to demote Pluto, those who desired to preserve Pluto's planethood, and those whose opinions could not be divined ahead of time. Marsden found himself supporting Stern's roundness criterion, at least initially.

The Working Group debated the question of what should be a planet ad nauseum through numerous e-mails sent during March and

April 2005. Every possible argument, constraint, or consideration regarding what should be a planet was presented, seconded, discussed, disregarded, or discarded, with no clear consensus emerging. The members tended to have strong opinions on this subject, and they were generally unwilling to change their opinions on the basis of arguments presented by others. Achieving a consensus among them was about as hard as trying to herd a group of 19 feral cats into a room with several open doors and windows. Eventually the Working Group tired of its hopeless task, and the e-mails stopped.

Then along came Michael Brown and Xena in July 2005, demanding that the IAU decide just what Xena/2003 UB313 was. The e-mail traffic picked up again. The chair of the Working Group, Iwan Williams of Queen Mary University of London, insisted that we make up our collective mind. The IAU Executive Committee had learned of our group's dithering and were threatening to take action themselves to settle the issue.

In August, Williams declared that we would take a vote. The results came in with a split decision: 7 in favor of keeping Pluto as a planet, as well as anything larger; 7 in favor of demoting Pluto; and 7 in favor of a compromise. Nineteen members could vote, yet a total of 21 votes were tallied; it was looking like a Chicago Democratic election. I had proposed as a compromise that we create several different categories: the eight major planets, the Kuiper Belt planets (Pluto, Xena/2003 UB313, and others), and the historical planets (the major planets plus Pluto). Given the vote, it was clear that the Working Group was not particularly in favor of a compromise.

The stalemate persisted throughout the next several months, as votes on three different compromises led to outcomes of 8 to 11, 11 to 8, and 6 to 12. Williams decided to declare an end to hostilities and to abandon the field by summarizing the three compromises the Working Group had voted on and sending his report to the IAU Executive

Committee in November 2005. Let *them* waste their time on this Gordian knot.

The IAU Executive Committee mulled over the Working Group's final report for a few months and then, in April 2006, decided to appoint a new committee to waste its time in trying to seek a consensus definition. This time the committee was small: just five astronomers, popular science writer Dava Sobel, and the chair, astronomy historian Owen Gingrich of the Harvard-Smithsonian Center for Astrophysics. Iwan Williams agreed to work again on this impossible task. The new Planet Definition Committee went to work, knowing that they had to come up with something in time for the upcoming IAU General Assembly in Prague.

After meeting at the Paris Observatory on June 30–31, the seven-member group quickly decided on a definition they could all support: Stern's roundness definition. They introduced their proposed resolution at the IAU's Prague meeting on August 16, accompanied by an article by Dava Sobel published in the *Washington Post* that same day, buttressing their case. Clearly they had planned the release of their decision well ahead of time and had gathered supporters, such as Neil deGrasse Tyson, director of the Hayden Planetarium in New York City, who applauded their decision. Richard Binzel, an MIT planetary scientist who was one of the Magnificent Seven on the Planet Definition Committee, predicted that their proposal would be approved by a wide margin in Prague. The question of Pluto would be settled once and for all. Pluto was a planet.

The August 17 issue of *Nature* contained a story about the IAU resolution with a differing opinion. Without knowing the details, I had predicted that a roundness criterion would be met with "a long line of people waiting for the microphone to denounce it." Reading all those endless e-mails from the Working Group had disillusioned me about the likelihood of easily achieving consensus on Pluto.

The Pluto proposal hit the IAU astronomers like a rotten melon. They could not conceal their disappointment in the proposal to retain Pluto as a planet—and to add dozens of other small, round bodies, including Ceres, to the hallowed list of planets. Each time the resolution and its evolving variations were presented, long lines of astronomers did indeed march to the microphone to register their vigorous objections. Several straw votes were held in these open sessions, and Pluto was shot down in flames each time. Owen Gingrich and his committee realized that they were facing a rejection of their Pluto proposal when it was brought up for a final vote on August 24. Admirably, Gingrich and his committee changed their course and tried to draft a proposal that stood a chance of being approved.

The next version they proposed on August 22 also failed. Time was short. What would be presented on the last day, August 24?

The IAU Executive Committee decided to present not one but two resolutions on planethood for the final assembly in Prague, each with two parts, for a total of four votes. This was going to be interesting. The fate of the four resolutions was to be decided by a simple majority vote of the IAU members present in the plenary hall.

Resolution 5A stated in essence that yes, planets had to be round, but they also had to be massive enough to have cleared their surroundings of any competing pretenders to planethood. Given that Pluto had lots of similar-sized siblings in the Kuiper Belt, Pluto would not be a planet if Resolution 5A passed, and it did in a landslide. Yellow cards were everywhere, and there was no need to try to count the cards in favor and against. Resolution 5B would insert the word *classical* in front of the word *planet* in Resolution 5A, effectively leaving Pluto a planet, but not one of the eight planets discovered by classical times (the last planet discovered before Pluto was Neptune in 1847). Resolution 5B was voted down by a margin of about 3 to 1. General applause arose once Resolution 5B was pronounced dead on arrival. Pluto was no longer a planet after that second vote. Now there were only eight.

Resolution 6A proposed that Pluto should be called a "dwarf planet," and this resolution passed by a vote of 237 to 157. Resolution 6B said that dwarf planets in the Kuiper Belt should be called "plutonian objects." Plutonian objects? "Plutonian objects" was offered as an improvement on the previous suggestion of "plutons," which, it turned out, had been claimed long ago by geologists as a word meaning a large underground body of once-molten rock, to the dismay of the astronomers who learned this bit of geological terminology only after making their "pluton" suggestion known to the entire world, thereby managing to enrage otherwise uninterested geologists and enlist their support in the battle against Pluto and the Plutons. Resolution 6B failed on the closest vote of the four: 183 for to 186 against. Pluto was now a dwarf planet.

As soon as the final vote was cast in Prague, my office telephone began to ring. I had been watching the vote through streaming live video over the web, having returned from Prague a few days earlier. The Pluto story had captured the attention of the media, and the outcome was a shocker. I spent the rest of August 24 being driven around Washington, D.C., in the sleek black limousines normally reserved for political figures and commentators, doing live television interviews for CNN and other television networks, talking about the demise of Pluto to the amusement of network anchors who are more accustomed to dealing with the demise of large numbers of human beings. The culmination of the day was a Red Top taxi ride (public television does it on the cheap, just like astronomers) to the studios of WETA in Northern Virginia to do "The NewsHour" with Jim Lehrer. Ray Suarez enjoyed the chance to talk with me under circumstances where we could laugh together about human foibles, and about the foibles of astronomers in particular. We ended the segment chuckling about what the new acronym should be that would help school children memorize the names and heliocentric ordering of the Solar System's eight planets.

August 25, 2006—Not everyone was laughing after the Prague votes. Alan Stern and Mark Sykes, director of the Planetary Science Institute in Tucson, Arizona, and a co-investigator on NASA's Dawn Mission to Ceres and Vesta, launched a petition drive protesting the IAU's decision. Within a few days, they had garnered the support of 400 irate planetary scientists, students, and engineers and were planning to bring up the question of Pluto again at the next IAU General Assembly, scheduled for 2009 in Rio de Janeiro, Brazil. Ceres had been provisionally promoted to full planethood for a little over a week in Prague, but now Ceres was merely a dwarf planet, along with Pluto and several others, while Vesta remained a lowly minor planet. It was an outrageous turn of events for the Dawn Mission team.

September 13, 2006—If Pluto was now a dwarf planet, so was Xena. Michael Brown was thus free to suggest proper names for Xena and Gabrielle. He suggested Eris and Dysnomia. Eris was the Greek goddess of discord and strife, and Dysnomia was her daughter. Angered at not being invited to attend a wedding, Eris had started a quarrel among the goddesses that led to the Trojan War. Brown had cleverly chosen names that were entirely appropriate for the two heavenly bodies that helped to launch the take-no-prisoners Battle of Prague. The relevant IAU nomenclature committees, tired by now of discord and strife, and pleased by Brown's good-natured approach to

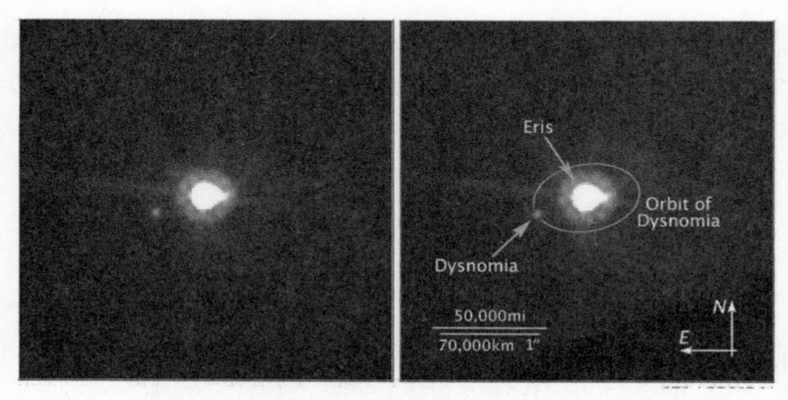

FIGURE 26. Eris and its satellite Dysnomia, as observed by the Hubble Space Telescope. [Courtesy of NASA, ESA, and Michael Brown (Caltech).]

the loss of planethood for Xena/2003 UB313, readily approved Brown's two suggested names.

Marsden assigned Eris the number 134340 on the Minor Planet Center's list of Solar System non-planets, and Pluto was assigned 136199—numbers that look like those held up by criminals in their mug shots. Pluto had entered the dwarf planet penitentiary with a life sentence, no doubt hoping to be sprung after the next IAU General Assembly in 2009.

September 22, 2006—Michael Griffin appointed nine new members of the NASA Advisory Council, the group charged with offering expert outside advice to the NASA administrator. The new appointments were necessary because several members had resigned a few weeks earlier. Two of the resignations (those of Wes Huntress and Eugene Levy, provost of Rice University) were tendered only after being requested by Griffin. Levy had a long history of involvement in NASA's plans to fly planet-hunting space telescopes, and as a former head of NASA's Office of Space Science, Huntress could not be happy with what was happening to the science programs that he had helped to create and nurture. Levy and Huntress had thought that their job on the Council was to offer advice to Griffin, but Griffin apparently wanted them to tell him which science programs should be cut or dropped. They resigned instead.

In their places, Griffin appointed the nine new members of the Advisory Council, charged with providing Griffin with "the best advice possible," which in practice meant the advice that Griffin wanted to receive. One of the new members was Alan Stern, now executive director of the Space Science and Engineering Division at the Southwest Research Institute. One wondered if an appeal of Pluto's life sentence might appear on the agenda for the next Council meeting.

October 26, 2006—The Kepler Mission held a science team meeting at the Lowell Observatory in Flagstaff, in a building just a short distance away from the telescope that Clyde Tombaugh had used to discover Pluto 76 years before. The Kepler science team had known from the previous meeting in April that Kepler was in trouble. The seemingly inevitable mission cost overruns had led JPL to replan the entire mission to minimize the impact on Kepler's science. Fabrication of the primary mirror was behind schedule, which was likely to result in a further delay of the launch, which was originally planned for 2006.

At Lowell we were relieved to learn that Kepler had passed its critical design review with excellent grades the previous month. The review is exactly as critical as its name states; it is one of several distinct hoops that any NASA mission must jump through or risk being subjected to an also aptly named Cancellation Review. Few missions survive the latter, although Kepler's Discovery Mission sibling, the Dawn Mission, was one example. Dawn was a true NASA zombie, having been terminated the previous March as a result of a Cancellation Review. Cost overruns were to blame, along with technical problems. However, the JPL engineers managed to resurrect Dawn a few weeks later by proposing solutions that NASA headquarters found palatable for addressing the technical and cost problems. We were all relieved to hear that Kepler would not have to undergo the near-death experience through which our sibling mission had suffered.

There was more good news for the team at Lowell. Kepler's primary mirror had been completed and delivered to the primary contractor, Ball Aerospace & Technologies Corporation in Boulder. Ball Aerospace, oddly enough, is a division of the same company that is famous for making Ball glass jars, rubber seals, and metal screw-on caps used for canning fruits and vegetables in home kitchens.

However, the JPL replan of Kepler had led to a slip in the launch date to November 1, 2008, and it didn't solve the cost overrun problems. Given the high grades on the critical design review, the Kepler team decided to ask NASA headquarters for more funds, in spite of the dire financial situation for NASA science. Under the replan, Kepler would cost a total of $568 million, well above the original cost cap for Discovery Missions of $299 million, although some of this cost growth had been a result of NASA headquarters' insistence that JPL run the project, rather than NASA Ames, where Borucki and most of the Kepler team worked. NASA headquarters was being asked for an extra $67 million to support the replan, and Mary Cleave had said to proceed with the new plan. Kepler seemed to be surviving reasonably well in the shark-filled waters of NASA's SMD.

We also learned that the competition, CoRoT, was nearly finished and was scheduled for launch on December 21, just a few months away. It was clear that CoRoT would have a head start of almost 2 years on Kepler. Confederate cavalry leader Nathan Bedford Forrest's advice for winning battles was "to git thar fust with the most men." CoRoT would "git thar fust," but Kepler's plan was to outperform CoRoT with a much larger space telescope and a mission designed to do one thing only, and to do it well: find Earths.

October 31, 2006—Michael Griffin announced NASA's plan to use the Space Shuttle to perform Hubble SM4 and enable Hubble to function for another decade. The repair mission would fly in May 2008 at a cost of $350 million. NASA felt that the risk of the shuttle mission to Hubble was worth taking, especially now that the shuttles had resumed space flights and returned safely three times since the Columbia reentry disaster in 2003. Senator Mikulski predictably exclaimed, "It's a great day for science." Space Telescope Science Institute astronomers were ecstatic.

It was a great day for Maryland science, but the rest of NASA's science program would have to pay the cost of the decision to keep Hubble alive. Nearly all of NASA's astronomy budget would be needed to support just three programs: Hubble, Webb, and a flying infrared telescope, the Stratospheric Observatory for Infrared Astronomy (SOFIA). There was no room for anything else, at least until the annual costs for Webb began to decline some time after 2008. Work on SIM had been halted earlier in the year to cover cost overruns on SOFIA. Planet hunting seemed to have disappeared from NASA's agenda, with the exception of the Kepler Mission. It was looking as though NASA headquarters might decide to rename Kepler TPF-K and forget about TPF-C and TPF-I altogether.

NASA was in the unsettling position of having to perform battlefield triage, choosing those missions that might live and leaving behind those that were expected to die.

November 15, 2006—The CoRoT spacecraft arrived safely at the Yubileiny airport, close to the Baikonour launch site in Kazakhstan, the primary launch site for the Soviet Union's and now Russia's commercial and cosmonaut rockets. On December 27, CoRoT was launched successfully into a circular, polar orbit around Earth on a Soyuz-2-1b rocket with a Fregat upper stage. CoRoT was launched 6 days late as a result of minor delays in getting the Soyuz rocket ready. With a total length of almost 14 feet (4.1 meters) and a launch weight of 1400 pounds (630 kilograms), CoRoT is basically the size of a Volkswagen bug.

The science team expected CoRoT to discover dozens of hot super-Earths, as well as equal numbers of hot Jupiters. Now that CoRoT was safely in orbit, it was time to see whether their high expectations would be met. How many super-Earths would CoRoT find? Might CoRoT find an Earth or two as well?

CHAPTER 7

The Comedy Central Approach

"There is something fascinating about science. One gets such wholesale returns of conjecture out of such a trifling investment of fact."

—MARK TWAIN (1835–1910)

January 2, 2007—At about 4:30 p.m. something smashed a hole in the roof of a house in Freehold Township, New Jersey. The object shattered some floor tiles in the upstairs bathroom, bounced back up and careened off the wall, leaving a dent, and then came to rest behind the toilet. An inhabitant of the house heard the sharp noise and thought it was a firecracker set off by somebody who had one left over from New Year's Eve celebrations. It was not until later that night that someone walked into the bathroom and discovered the damage. During the clean-up of the mess of shredded fiberglass insulation and dry wall fragments, the culprit was found hiding behind the toilet: a crescent-shaped, silvery-brown object about 4 inches long. The alignment of the broken floor tiles and the hole in the roof showed that the object had struck the house at an angle very close to the vertical: it had come almost straight down from the sky. But what was it? The Nageswaran family suspected it was a piece of either an airplane or a satellite—or maybe a meteorite, because the object did not look like anything they had seen on Earth.

The next morning the family called the Freehold Township police to report the intrusion. The police arrived shortly thereafter and took the suspect into custody. At the police station, the officers left the object in their cruiser in case it was radioactive. A quick check with a Geiger counter showed that it was not, and the object was taken inside and placed in an evidence container. Investigators from the Federal Aviation Administration (FAA) arrived a few hours later and, after several hours of study, declared that the object showed no signs of having been manufactured on Earth and hence had not been derived from an airplane. The FAA investigators departed.

The story had got out that the Freehold police were onto something interesting, and television stations sent crews to film the events at the police station for the evening news. The next day several scientists arrived to investigate the object, including Jeremy Delaney of Rutgers University, an expert on meteorites. The scientists scrutinized the object as best they could, measuring its density and magnetic properties, and comparing its surface to those of known meteorites, and they finally determined that it was probably an iron meteorite. The literal acid test would be to saw the object in half and look for the characteristic "Widmanstatten patterns" that appear on the sawed faces when some iron meteorites are etched with acid. However, the Nageswaran family did not want to have their meteorite sawed in half for fear that it's value would be decreased as a result.

A meteorite that had struck and passed through the trunk of a Chevy Malibu in Peekskill, New York, on October 9, 1992, had earned the owners a pretty penny when they sold both the meteorite and the Malibu. The meteorite had entered Earth's atmosphere on a Friday night during high school football season, with the result that dozens of alert football fans, recording their sons' immortal exploits with portable video cameras, swung their cameras from the well-lit playing fields to the skies to capture the vivid streaks of bright green light that had suddenly appeared. The parent meteoroid had broken into multiple fragments whose incandescent wakes in the upper atmosphere outshone the stadium lights as the fragments slammed deeper and deeper into Earth's thin veneer of oxygen gas, producing the green light. George Wetherill and Peter Brown of the University of Western Ontario, Canada, had spent the better part of a year gathering these videotapes, going to the high school stadiums where the games were played, and using a surveyor's theodalite to plot the trajectory of the meteoroid on the sky, based on the location of the light poles and other stadium features seen on the videos. They were able to determine the orbit of the meteroid that had produced the Peekskill meteorite, and it was only the fourth time that an orbit had been determined for a found meteorite. They estimated that the parent meteroid was about 3 feet in diameter, weighed several tons, had a mean orbital distance from the Sun of 1.5 AU, and orbited every 1.8 years outward to the inner edge of the main asteroid belt from which it had originated.

The owners of the Peekskill meterorite and the Chevy Malibu put them both up for auction and sold them for $75,000 and $25,000, respectively. Although the Nageswarans decided not to try to auction off their slightly damaged house for the price of a Chevy Malibu with a hole in the trunk, they did think that they had a potentially valuable meteorite. The Nageswarans' meteorite was put on display for a day at the Rutgers University Geology Museum, unblemished by further scientific investigation.

January 4, 2007—Around 6:45 a.m., traffic reporters flying in helicopters in the area of Denver, Colorado, trained their live video cameras on a dozen bright streaks of light that had suddenly appeared in the sky over the Rocky Mountains. The streaks came from the north and were headed south, toward Colorado Springs, the home of Peterson Air Force Base and the U.S. North American Aerospace Defense Command (NORAD). NORAD, the Cold War entity created in anticipation of a nuclear war with the Soviet Union, had built its doomsday missile command center deep within Cheyenne Mountain, one of the Rocky Mountains near Pike's Peak. The spectacular fireworks display was similar to what the end game might have been for the Cold War: multiple, independently targeted warheads screaming down from space. Cheyenne Mountain would have been at the top of the Soviet hit list for a first or retaliatory strike on the United States.

A spokesman from Peterson Air Force Base later explained that the ominous threats seen over the Rockies were portions of the upper stage of the Russian rocket that had launched a French satellite into Earth orbit a week earlier. That could mean only one thing: CoRoT's Russian launch vehicle had struck the American heartland. The old Soviet ICBM had faithfully delivered CoRoT to the proper orbit, but the upper stage of the ICBM disintegrated in a fireball over Colorado. The Russian rocket nearly managed to carry out its Dr. Strangelovian mission long after the Cold War ended.

After the public display of the New Jersey meteorite in January, the Nageswarans decided to have their object tested further in order to remove any doubt about its authenticity. They hoped to sell it to a museum so that it could be viewed by the public, perhaps stimulating a greater interest in science—as well as making them a tidy profit. Jeremy Delaney told the Nageswarans that the object could be examined with a scanning electron microscope that had recently been installed at the American Museum of Natural History in New York City. The microscope could determine the composition of the object without its having to be sawed in half. Instead, the microscope would harmlessly fire electrons at the object, causing the surface to emit X-rays. The microscope could then detect the X-rays and determine what elements had emitted the radiation. The result would be a map of the surface of the object, showing the location and concentration of the most common elements in iron meteorites, namely iron and nickel.

Denton Ebel, curator of the museum's meteorite collection, took charge of the analysis of the Nageswarans' object. The results came back quickly and definitively. The object was indeed mostly iron, as expected, but it had no nickel. No nickel meant it was not an iron meteorite, at least not one like any other iron meteorite that had ever been found. The electron microscope had also shown that the iron contained a small amount of chromium and manganese, which meant that it was in fact a 13-ounce lump of stainless steel. Mother Nature does not make stainless steel; human beings do. The Nageswarans' meteorite had turned out to be a piece of space junk, not a meteorite. But what kind of space junk? American space satellites do not use stainless steel, but rather titanium. That meant the Russian space program was likely to be the source of the object, because the American and Russian space programs were the primary polluters of the skies (at least before the Chinese used a warhead to destroy one of their own failed weather satellites later in 2007). The fact that the object had fallen in New Jersey just two days before pieces of the old Soviet ICBM had fallen over Cheyenne Mountain was a powerful clue about its probable origin. The Nageswarans' house may have been struck by an errant piece of CoRoT's ticket to low Earth orbit.

January 27, 2007—The Advanced Camera for Surveys (ACS) on Hubble suffered an electrical failure and went blind. The camera was the newest instrument on Hubble, having been installed in the fourth Hubble servicing mission in 2002. The camera had twice the field of view of Hubble's previous wide-field camera and could collect data 10 times faster. As a result, it quickly became the most popular of the four instruments still working on Hubble. Its loss was a major blow to Hubble science, and astronomers were left to wonder whether the magicians at NASA's Goddard and the Space Telescope Science Institute would be able to bring ACS back to life again by cycling the power and trying other tricks. The engineers were pessimistic and doubted that anything could be done to revive the camera, short of the hands-on attention of NASA astronauts with new parts. The plan to install a new wide-field camera on the fifth servicing mission, in 2008, would mean that some projects could be carried out with the new camera, though several years later than planned. Hubble was aging even faster than they had feared.

February 1, 2007—A Japanese planet-hunting team led by Bun'ei Sato of the Okayama Astrophysical Observatory in Japan had a paper accepted for publication in the *Astrophysical Journal*. Although the Japanese have their own journals for astronomical publications, many realize that if they want U.S. astronomers to read their papers and cite their work, it is better to publish their results in U.S. journals.

Sato and his group had developed their own iodine cell for use with a spectrometer on the 75-inch (1.88-meter) telescope at the Okayama Astrophysical Observatory located on Mt. Chikurin-Ji, the best astronomical site in Japan. The 75-inch telescope is also the largest telescope in Japan, though Japanese astronomers have built their own large telescope on Mauna Kea in Hawaii, the 27.3-foot (8.2-meter) Subaru Telescope, located next to the twin Keck Telescopes, each with a diameter of 33.3 feet (10 meters), and close to the U.S. National Optical Astronomy Observatory's 27.0-foot (8.1-meter) Gemini Telescope. Subaru is thus a smidgen larger than Gemini, giving it the bragging rights as the next largest telescope on a premier observing site after the Kecks. The telescope was named Subaru after the Japanese word for the Pleiades, a cluster of thousands of young stars in the Taurus constellation, as was the Japanese Subaru automobile company, whose logo features six stars.

Telescope time is just as precious on Subaru as on any other large telescope, and planet hunting requires a lot of telescope time, so Sato and his group launched their planet search on the Japanese 75-inch telescope in 2003. They decided to focus on a type of star that had been largely ignored by the other Doppler search groups, namely giant stars. Giant stars have lived long enough to have begun burning elements other than hydrogen in their centers (helium, carbon, and oxygen), and as a result they swell in size by factors of 10 to 100. A red giant star with a diameter 100 times larger than the Sun would engulf any hot Jupiters that had been in orbit around it, but more distant planets could be expected to survive, although they might well become hot Jupiters themselves as a result of the much higher luminosity of the giant star. Any planets still left orbiting giant stars would have orbital periods of a year or more, so the planet hunters would have to be patient; one could not expect to find any planets with 4-day orbital periods around a giant star.

Sato and his team searched giant stars in the Hyades open cluster for planets. The Hyades is the closest open cluster of stars to the Sun, about 140 light-years away, and it consists of some 400 stars about 625 million years old. The Hyades are easily seen on a winter night in the Northern Hemisphere as a V-shaped group of stars in the Taurus constellation. Bill Cochran and his Texas colleagues had searched the Sun-like stars in the Hyades for hot Jupiters back in 2002, but they had come up empty-handed.

By 2006 Sato's team had found one giant star named Epsilon Tauri that had a planet with a mass of at least 8 Jupiter masses orbiting Epsilon Tauri every 594 days, at a distance almost twice that of Earth from the Sun, 1.9 AU. Epsilon Tauri's mass was estimated to be about 2.7 times the mass of the Sun, making it the world heavyweight champion for stars with planets.

Considering that Cochran had found nothing in the Hyades dwarf stars, Sato's detection hinted that more massive stars might have even more gas giant planets than lower-mass stars like the Sun. This hint became stronger when two other groups submitted papers for publication in February 2007. The California-Carnegie planet search team submitted a paper to the *Astrophysical Journal* with the first results from their Doppler survey of intermediate-mass stars, those with masses as much as twice that of the Sun. They had found three intermediate-mass stars with planets, making a total of nine known planets around stars with masses at least 60% greater than that of the Sun. The Geneva Observatory group similarly submitted a paper to *Astronomy & Astrophysics* that month, reporting their detection of a planet orbiting a star with a mass 2.4 times that of the Sun. They concluded that more massive stars have more planets than less massive stars, probably because more massive stars have more massive planet-forming disks.

Given the high frequency of planets found around Sun-like stars, the fact that higher-mass stars seem to have even more planets was highly encouraging. The planets orbiting giant stars are all gas giants, but their existence means that super-Earths and even smaller rocky planets are almost certainly in orbit as well, waiting to be discovered.

February 21, 2007—NASA headquarters once again had the enviable problem of an abundance of riches. Three different teams of astronomers had used Spitzer to try to do the same thing: measure the light emitted by an extrasolar planet at a range of infrared wavelengths. If successful, these three teams would share the honor of having measured the first spectrum of the atmosphere of a planet outside the Solar System. Being able to measure the light given off at different wavelengths by a planet's atmosphere is our best hope for determining whether the planet is habitable—or maybe even inhabited. The Terrestrial Planet Finder would be expected to search for the presence of molecules of carbon dioxide, water, oxygen (or ozone), and perhaps methane in order to make the case for habitability and even life. The three teams had taken the next major step in this direction by using Spitzer to search for evidence of water in the atmospheres of two different hot Jupiters.

Because Spitzer is a NASA space telescope, NASA headquarters had the power to ensure that all three teams received more or less equal credit for their discoveries, even though their results would not be published simultaneously (order of submittal for publication being the usual criterion for establishing priority in science). One of the teams had a paper about its results scheduled for publication by *Nature* on February 22, 2007, so NASA headquarters decided to hold a telephone conference for the press on the day before to announce the results of all three teams.

At 1:00 p.m. Eastern time on February 21, Jeremy Richardson of NASA Goddard led off with his team's discovery. Richardson's group had used Spitzer's Infrared Spectrograph to measure the infrared spectrum of the first transiting planet, the hot Jupiter orbiting the star HD 209458. The idea was to observe HD 209458 b not when it passed in front of the star, but during the secondary eclipse, when the planet passed behind the star. By measuring the amount

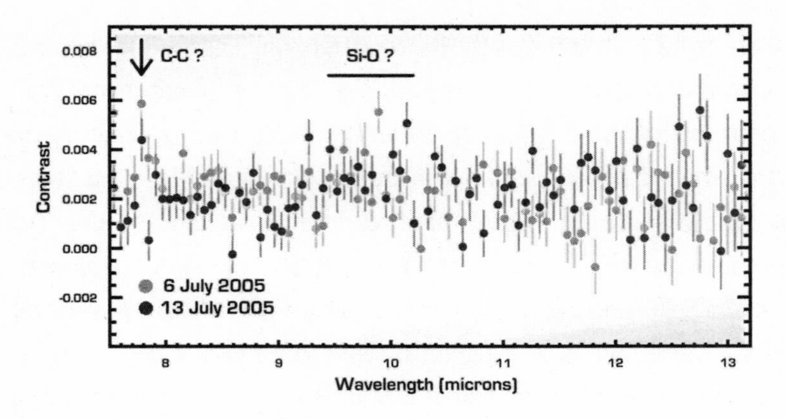

FIGURE 27. First spectrum of an extrasolar planet, HD 209458 b, as observed by the Spitzer Space Telescope, showing the light emitted by the planet at mid-infrared wavelengths. [Courtesy of NASA, JPL, Caltech, and Jeremy Richardson (NASA GSFC).]

of light emitted by the star and the planet just before the secondary eclipse occurred, and then subtracting the light emitted by the star alone during the secondary eclipse, Spitzer could estimate the amount of light that had come from the planet. Spitzer had already accomplished this feat for HD 209458 b and TrES-1 b, but those detections had been accomplished by monitoring changes in the amount of infrared light emitted by the two hot Jupiters over much broader ranges of infrared wavelengths than Spitzer was being asked to work with now. This was a tough problem, even for Spitzer, which had not been designed to measure the spectra of extrasolar planets.

By using Spitzer's infrared spectograph to measure the light emitted by the planet at wavelengths between 7.5 and 13 microns (1 micron = 0.00004 inch), Richardson could search for evidence of water vapor in the hot atmosphere of HD 209458 b. The prediction was that the spectrum would show less light being emitted at the shorter wavelengths than at the longer wavelengths, because water molecules were expected to absorb more of the planet's light at the shorter wavelengths. Spitzer produced a spectrum that was pretty much flat across the wavelength range of 7.5 to 13 microns. Apparently HD 209458 b had no water in its atmosphere, which was puzzling, because gas giant planets are thought to be composed of a fair amount of water gas and ice, regardless of how they are formed. Maybe HD 209458 b was just an oddball?

The next person up to bat at the press telecon was Carl Grillmair of the Spitzer Science Center in Pasadena, California, where Spitzer is managed and operated. The astronomers who work at the Spitzer Center love to get a chance to use Spitzer to do their own science, just as the astronomers at the Space Telescope Science Institute love to use Hubble for their own observations. Telescope time on both Spitzer and Hubble is awarded competitively, so if the Spitzer Science Center or Space Telescope Science Institute astronomers write the best proposals for using Spitzer or the Hubble telescope, they get the time. And then they often reap the the reward of having to work overtime in order to analyze their precious observational results, in addition to performing their normal day jobs keeping the two space telescopes working properly.

Grillmair and his colleagues had used Spitzer to do the same thing as Richardson's group, but for a different hot Jupiter: HD 189733 b. This hot Jupiter had been discovered by the Geneva Observatory's Doppler search program in 2005 and, afterward, was discovered to be a transiting planet. Because HD 189733 b's orbital period was 2.2 days, shorter than HD 209458 b's 3.5 days, HD 189733 b orbited closer to its star and so was likely to be hotter. Infrared light from a planet is thermal emission, and hotter planets give off more infrared light. In addition, HD 189733 b's star is a K dwarf, with lower mass and hence a smaller radius than HD 209458 b's star, a G dwarf like the Sun. As a result of these two factors, the detectability of infrared light from HD 189733's planet was about twice as good as that of infrared light from HD 209458's planet, so Grillmair and his team decided to study the former rather than the latter.

Spitzer followed the secondary eclipses of HD 189733 b's planet in October and November 2006. Grillmair found that the infrared spectrum of HD 189733 b looked much like that of HD 209458 b: relatively flat and featureless. Once again there was no hint of absorption at shorter wavelengths caused by water vapor in the planet's atmosphere. Was it just not there, or had the two teams missed something in their analysis?

The third speaker was Mark Swain of JPL. His team had not analyzed any new Spitzer observations but, rather, had re-analyzed the same Spitzer data for HD 209458 b that Richardson's team had analyzed. Spitzer had performed the HD 209458 b observations in July 2005, and the data were duly entered a little over a year later into the Spitzer Space Telescope archives, where they could be accessed by anyone with the time and interest to do so. As a government agency, NASA takes seriously its responsibility to give the public direct access to the data taken by its space telescopes, after the astronomers who requested the data have had a chance to digest the results and get a research paper submitted for publication. The proprietary period for Spitzer data is 1 year from the time that the proposers receive the processed data. Thereafter, the data are fair game for anyone. Richardson's team had been a bit tardy in getting their analysis of the HD 209458 b data completed and submitted to *Nature*, which gave Swain and his team a chance to beat them to the punch.

Swain and his group performed two different analyses of the data for HD 209458 b and came up with the same answer both times: there was no evidence of water in the planet's infrared spectrum. Swain was thus able to confirm Richardson's result for the same planet. Grillmair had shown that the apparent absence of water absorption was not limited to a single hot Jupiter: two hot Jupiters exhibited the same flat infrared spectrum.

As the outside commentator during the telecon, I had made the point that NASA's game plan for searching for life on Mars was to "follow the water" and that the same game plan held for extrasolar worlds. The three teams had used Spitzer to do the obvious: search for water on two hot Jupiters. The fact that all three teams were in complete agreement that HD 209458 b and HD 189733 b showed no evidence of water vapor made a convincing case that the Spitzer data had been properly reduced and analyzed. The only problem was that the final answer was not the one we had been expecting or wanting. Where was the water?

April 2, 2007—Mary Cleave took early retirement from her job running the Science Mission Directorate at NASA headquarters. Michael Griffin decided to replace Cleave with a genuine space scientist, one who had long experience not only with the science side of the house but with the space mission side as well. Griffin needed to look no further than his own NASA Advisory Council to find the perfect replacement for Cleave: Alan Stern. With a Ph.D. in astrophysics and planetary science, extensive publications on subjects spanning the space sciences, and leadership of the ongoing Pluto Mission, Stern appeared, if anything, to be overqualified for the job. But how would he handle the seemingly impossible task of managing an annual science budget of $5.4 billion that, however huge it seemed, was still a billion or two short of what U.S. scientists had long been expecting?

Stern took the job with the understanding that no new funds would flow to SMD. If something new were to be started, something else would have to be stopped. Stern acknowledged that there would be things he had to do that would cause pain. In fact, Stern would need to change the culture that led to cost overruns on practically every NASA space mission. NASA contractors had grown accustomed to making a bundle on their space missions. (European companies, by contrast, tended to view their involvement in such projects as a badge of national pride and so worked on fixed-price contracts.) Whenever a NASA mission ran into trouble, the standard solution was to go to NASA headquarters and get more money. The overruns on the Webb Telescope alone were enough to pay for SIM.

On the same day that Alan Stern took over the reins for NASA science programs, the CoRoT telescope began routine astronomical operations, after having been successfully tested in orbit for several months. CoRoT's first field of stars was in the constellation Monoceros, where CoRoT would stare for 60 days before moving on. The next two fields would be toward the center of our Milky Way galaxy, with first a short 26-day campaign and then a long 150-day campaign planned. CoRoT was on the march, directed by its generals at the mission control center in Toulouse, France.

April 4, 2007—Stephane Udry and his colleagues on the Geneva Observatory team submitted a paper to *Astronomy & Astrophysics* reporting the Doppler detection of two more planets around the M dwarf Gliese 581. In 2005 the Swiss team had found a "hot Neptune" with a mass of at least 16 Earth masses orbiting Gliese 581 with a period of 5.3 days. They had continued to monitor Gliese

581's wobble and now had evidence for two more planets. The minimum masses for the two new planets were 5 and 8 Earth masses, making for a planetary system with three super-Earths. The smallest of the three might even beat out the planets Gliese 876 b and OGLE 2005-BLG-390L b for the distinction of being the lowest-mass planet known around a star.

The two new super-Earths orbited with periods of 13 and 83 days, indicating that they were located far enough away from their M dwarf star to be warm and cool super-Earths, respectively. The 5-Earth-mass planet was orbiting at the inner edge of the expected habitable zone for an M dwarf with one-third the mass of the Sun, whereas the 8-Earth-mass planet orbited close to the outer edge of the habitable zone. If their orbits were noncircular, they would spend at least part of their time inside the habitable zone, where liquid water could exist at their surfaces.

The Swiss team had found a three-planet system with one or possibly two habitable planets. Their discovery was made possible by the development of a new, higher-precision spectrometer called HARPS: the High Accuracy Radial Velocity Planet Searcher. HARPS was installed on the 114-inch (3.6-meter) European Southern Observatory telescope at the La Silla Observatory in Chile. HARPS was able to achieve Doppler precisions on the order of 2.3 miles per hour (1 meter per second) or better. This precision was needed to detect the Doppler wobble induced by these two new planets, which was only about 7 miles per hour (3 meters per second) for both. This signal was 20 times smaller than that due to 51 Pegasus's planet. To keep the HARPS spectrograph stable and happy, it was locked into a climate-controlled vault below the dome of the La Silla telescope. The rumor was that only one key existed for the vault door, and it was kept in Geneva, so that there could be no chance of anyone in Chile accidentally disturbing HARPS.

April 8, 2007—Giovanna Tinetti of the European Space Agency in Frascati, Italy, and her colleagues submitted a paper to *Nature* claiming the detection of water vapor on the planet HD 189733 b, an outcome contrary to the results found by Carl Grillmair's group using Spitzer. Tinetti and her colleagues had also used Spitzer, but their analysis pertained to the primary transit, when the planet passed in front of the star, not to the secondary eclipse studied by Grillmair. They used the Infrared Array Camera on Spitzer to study what happened to the star's light as it passed through the atmosphere of the planet during the primary eclipse. Tinetti and her team found that the effective size of the planet (that is, the amount of the star's light being blocked out) depended on the wavelength of light coming from the star. The variation was what one would expect if the atmosphere of HD 189733 b contained water vapor, as had been expected prior to the news conference of February 21. Tinetti and her colleagues suggested that the failure of the other teams to detect water in HD 189733 b resulted from the planet's having an atmosphere with the same temperature throughout its upper layers. In such an atmosphere, there would be no way for water vapor to absorb some of the infrared emission coming from hotter layers lying farther down in the atmosphere, layers that might not exist if Tinetti was right. We were back to following the water.

Or were we? A month earlier, another European team, led by David Ehrenreich of the Institute of Astrophysics of Paris, had submitted a paper to the *Astrophysical Journal* analyzing exactly the same thing that Tinetti's group studied: Spitzer Infrared Array Camera observations of HD 189733 b's atmosphere during the primary transit. Ehrenreich and his team concluded that although they could use these observations to measure the radius of the planet to within 1%, this level of accuracy was not enough to permit a conclusive determination that the effective radius varied with the infrared wavelength in a manner consistent with the presence of water vapor. Ehrenreich's team included an astronomer who was also on the Tinetti team, Alfred Vidal-Madjar, who had used Hubble in 2003 to detect the presence of hot hydrogen gas escaping from HD 209458 b. Ehrenreich's paper, published three months after the Tinetti paper, was careful not to condemn the Tinetti paper, but it was clear that his team's analysis called Tinetti's claim for water vapor on HD 189733 b into question.

The scent had disappeared, and the trail of extrasolar water was lost again. The hunting dogs circled in confusion, barking and howling, waiting for fresh directions on how and where to renew the search.

April 23, 2007—The European Southern Observatory distributed a press release announcing the discovery of a habitable world around Gliese 581, singling out the smaller of the two new planets for this honor. Udry's paper was still under review at *Astronomy & Astrophysics,* but the Swiss team decided to go public with their find. The European Southern Observatory wanted to trumpet what could be done with HARPS, now the leading planet-finding spectrograph in the world, and the fact that HARPS had found a planet likely to be habitable was the ideal opportunity to blow their horn.

The European Southern Observatory press release made a splash around the world. I was in Lund, Sweden, at the time, helping to celebrate the first 25 years of the awarding of the Crafoord Prize. The Crafoord Prize is awarded by the Royal Swedish Academy of Science and is presented to the laureate by the King of Sweden, along with a sizable check. The Crafoord Prize is given in astronomy, mathematics, geosciences, and other fields that are not covered by the Nobel Prize categories. It is the astronomer's Nobel Prize. It happened that the King was unable to attend the prize ceremony in the

Lund Cathedral this time, but the Queen of Sweden, Silvia, did the presentation on his behalf to a geoscientist and a biologist, and afterward she led us to a banquet hall for a 5-hour feast of multiple courses and toasts to the Queen's health. After several hours of feasting and wine, those of us who were still able to stand up were ready to toast anyone's health.

The e-mails started rolling in as soon as the European Southern Observatory press release was sent out. I answered e-mail questions from American reporters about the Swiss discovery for most of the day. National Public Radio recorded an interview with me over the telephone in my hotel room, and the next morning many Americans woke up to hear me talking about the latest European exoplanet discovery on NPR's "Morning Edition." As a result of that radio exposure, a fresh round of e-mails arrived, including several requests for television interviews. This story had legs, in the lingo of the media.

A few days later, Stephen Colbert suggested, on his *Comedy Central* program "*The Colbert Report,*" that he thought the new habitable world would be a fine planet for the USA to take over after we finished trashing planet Earth and making it uninhabitable. What Colbert failed to realize, however (as did his astronomer guest, the American Museum of Natural History's Neil Tyson), was that Gliese 581's habitable planet was discovered not by Americans but by a team of Swiss, French, and Portuguese astronomers. By rights, the new world belonged to the European Union.

May 3, 2007—The first discovery of an extrasolar planet by the CoRoT space transit telescope was announced. The planet was named CoRoT-Exo-1 b, and it had a mass of 1.3 Jupiter masses with a 1.5-day orbital period. CoRoT had found its first hot Jupiter, the easiest prey to hunt down and catch. This hot Jupiter was a particularly

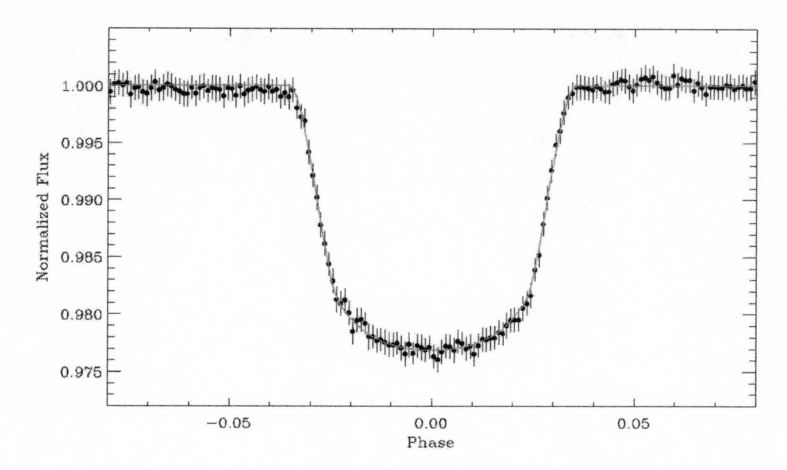

FIGURE 28. **First detection of a transiting planet by the CoRoT space telescope, showing the dimming produced by CoRoT-Exo-1 b, a hot Jupiter. [Courtesy of Pierre Barge (Astrophysics Laboratory, Marseille) and the CoRoT Mission.]**

fluffy gas giant, with a radius 1.65 times that of Jupiter, making it all the easier to find by the transit method.

CoRoT had drawn first blood in the space transit telescope battle. What would come next? The press release noted that CoRoT was functioning at least as well as predicted, and in some areas better than expected. In particular, CoRoT was able to measure stellar brightnesses to an accuracy of one part in 50,000 with enough observations, close to what the Kepler Mission was planning to achieve to find Earths. The CoRoT press release stated that CoRoT could detect "very small exoplanets similar to Earth," the grand prize of the field of extrasolar planet searches. The first team that found one might even get a Nobel Prize in physics, and not have to settle for a Crafoord Prize in astronomy. CoRoT was leading the way to Stockholm.

CHAPTER 8

Transits Gone Wild

I can imagine no discovery more fundamental than life on other planets, here in the Solar System, or around some other star.

—JOHN C. MATHER (OCTOBER 5, 2007)

May 6, 2007—Michael Gillon and his colleagues from the Geneva Observatory planet search team submitted a paper to *Astronomy & Astrophysics* Letters reporting the first detection of a transit by a hot super-Earth. Their detection proved that a hot super-Earth with a true mass 22.6 times that of Earth was orbiting Gliese 436 every 2.6 days.

By early 2007, about 10 hot super-Earths had been discovered by high-precision Doppler spectroscopy, but none had been found to transit. The odds were that for planets on such short-period orbits, like the hot Jupiters, 1 out of 10 should be observed to pass in front of their stars as seen from Earth and generate transit events. The first hot super-Earths had been detected in 2004, yet not one had been seen to transit. In a sense, the fact that no transits had been seen strengthened the case for rocky planets: if the hot super-Earths really did signal the existence of a large population of unseen rocky, Earth-like

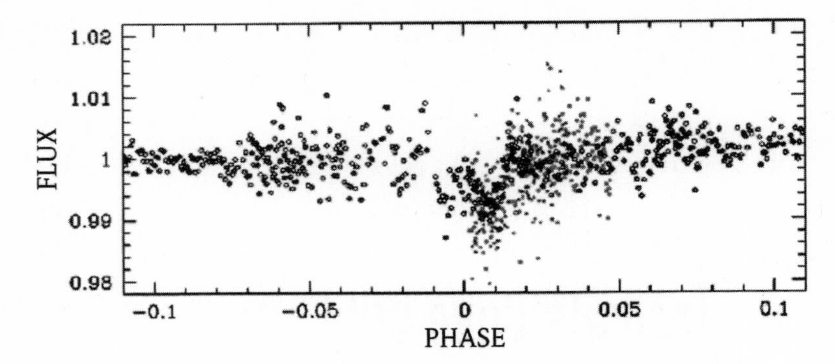

FIGURE 29. First detection of a transit by a super-Earth, GJ 436 b, as observed by several small ground-based telescopes. [Reprinted, by permission, from M. Gillon et al., 2007, *A&A*, volume 472, page L14. Copyright 2007 by the European Southern Observatory.]

planets, then their radii might be so small as to prevent detection of their transits by Earth-based telescopes. Such small planets would block out only a tiny fraction of their host star's light—a fraction too small to be seen by a telescope peering through Earth's turbulent atmosphere. If the hot super-Earths really were rocky planets, then we might have to wait for CoRoT or Kepler to catch the first transit of a hot super-Earth. This was one of those rare occasions when the absence of a detection seemed to be more promising than the detection of a transit by a hot super-Earth that turned out to have such a large radius that it could not be composed primarily of rock and even denser iron.

Gliese 436 b had been discovered by Paul Butler and the California-Carnegie planet search team in 2004. Gliese 436 b was the first member of a new class of planets, the hot super-Earths. Butler and his colleagues had checked to see whether they could detect a transit but were unable to find any dimming of Gliese 436. Gillon and his team nevertheless decided to add Gliese 436 to their list of transit survey targets, primarily because the low mass of Gliese 436, which was half that of the Sun, meant that Gliese 436 was considerably smaller than the Sun and so might show more easily evidence of a transit that could be detected from the ground.

Amazingly, Gillon was able to detect the transit of Gliese 436 b using the 24-inch (0.6-meter) telescope of the François Xavier Bagnoud Observatory in Switzerland, along with several telescopes of similar size in Chile and Israel that made confirming observations. The planet crossed the star close to the edge of the star's disk, making the transit much harder to detect than if it had sailed across the center. That explained why Butler's team had not been able to see a transit when they looked in 2003 and 2004.

The Swiss team had beaten CoRoT to the punch. They found that the radius of Gliese 436 b was about 4 times that of Earth, comparable to that of the ice giant planets Uranus and Neptune. Gillon and his colleagues concluded that Gliese 436 b's composition was likely to be dominated by water ice rather than by rock, so it should be called a hot Neptune rather than a hot super-Earth. In fact, given the heating from Gliese 436 that the planet had been experiencing for the several billion years of its existence, the water ice was likely to have melted, at least in its outer layers, turning Gliese 436 b into a "water world" worthy of the Kevin Costner movie.

Because Gliese 436 b seemed to be a hot Neptune instead of the hoped-for hot super-Earth, the question of how and where it formed arose immediately. The planet orbited at a distance from its star of 0.03 AU, a distance only 14 times greater than the radius of the star itself. Formation of an ice-rich planet that close to Gliese 436 was impossible, even for a star with half the mass of the Sun; ice would not be stable at the temperatures expected there. Presumably Gliese 436 b formed much farther out from its star and then migrated inward to its present orbit, much like the hot Jupiters seemed to have done.

However, there was a major problem with such a scenario, because the primary mechanism for planetary orbital migration is widely thought to be interactions of the planet with the gas of the planet-forming disk. If a hot Jupiter formed by either core accretion or disk instability, then it must have formed while the gas was still there; otherwise, it could not have become a gas giant. If the gas was still there, then it would still have a chance to migrate inward and become a hot Jupiter. However, if a Neptune-mass planet formed by core accretion while the gas was still there, so that it could migrate inward and become a hot Neptune, why didn't it gobble up enough of the disk gas to become a gas giant instead of remaining merely an ice giant? A 22-Earth-mass core should be able to quickly swallow enough cold outer-disk gas to become a Jupiter, yet this evidently did not happen to Gliese 436 b.

Forming a hot Neptune is equally prohibited in the disk instability scenario, where future ice giants must form at distances large enough for ultraviolet radiation to strip away the local disk gas and then denude the outer giant gaseous protoplanets, converting them into ice giants. If that were the case, the outer-disk gas would be lost before the cold Neptune formed, leaving it no obvious means for migrating inward to become a hot Neptune.

According to either giant planet formation theory, then, Gliese 436 b should not be an ice giant, at least not on a short-period orbit. What was going on here? There was also another fact to consider: the Doppler wobble of Gliese 436 showed a long-term trend of small amplitude, which implied the presence of a more massive third body in the system, orbiting at a much greater distance than Gliese 436 b. Perhaps Gliese 436 b had a long-period gas giant sibling. Such gas giant planets had been found orbiting at greater distances for the hot super-Earths around Gliese 876, 55 Cancri, and Mu Ara. If Gliese

436 had such a gas giant, Gliese 436 b was unlikely to have migrated inward past that massive sibling, and this suggested that Gliese 436 b had formed as an inner super-Earth rather than as an outer cold Neptune. Gliese 436 b as a hot Neptune just didn't make sense.

Whatever it was, Gliese 436 b had taken its time in revealing its true mass and size. Gliese 436 b was the first hot super-Earth detected and the first one found to transit, though only after 3 more years of searching and after another 10 hot super-Earths had been found. By contrast, HD 209458 b was the first hot Jupiter found to be transiting, but it was also the tenth hot Jupiter detected by a Doppler wobble, the opposite of the case for the hot super-Earths. Mother Nature seemed to enjoy playing tricks on us.

May 8, 2007—Luca Pasquini of the European Southern Observatory had a stunning result that promised to reverse much of the thinking about how gas giant planets formed. He and his European colleagues submitted a paper to *Astronomy & Astrophysics* suggesting that pollution might be the main cause of something called the metallicity correlation. This would be pollution on the scale of entire planets disappearing into the surfaces of their host stars, producing an oil slick of iron atoms across the entire surface of the star. Pasquini's discovery threatened to overturn the usual interpretation of the "metallicity correlation."

Fewer than a dozen extrasolar planets had been found by 1997, but that was enough for Guillermo Gonzalez, of the University of Washington in Seattle, to point out that the stars with known planets were different from the general population of stars that were being searched for Doppler wobbles. Gonzalez showed that the planet-host stars tended to have measurably more metals like iron in their spectra

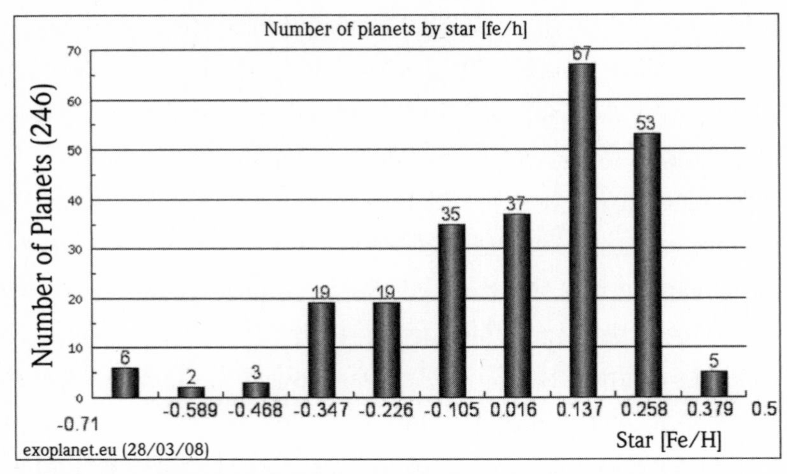

FIGURE 30. The metallicity correlation for extrasolar planets, showing that dwarf stars with higher amounts of iron relative to hydrogen ([Fe/H]) tend to have more planets discovered than stars with lower metallicity. [Courtesy of Jean Schneider, Paris Observatory.]

than the typical star in our neighborhood of the Galaxy. Iron serves as a proxy for all the elements in the periodic table heavier than hydrogen and helium, the two lightest elements. To an astronmomer, all such elements are "metals"; once iron is detected, it can be assumed that most of the rest of the periodic table is there as well. The metallicity correlation found by Gonzalez became stronger as more planets were found, and it soon became a determinant of which stars should be searched for hot Jupiters and the like; metal-poor stars were often dropped from planet search lists.

There were two distinct interpretations for what the metallicity correlation meant. First, the effect might be due to pollution of the outer layers of the planet-bearing stars if they had eaten some of their planetary progeny. If gas giant planets had to migrate inward to become hot Jupiters, perhaps some of them could not stop before being swallowed whole by their unforgiving mother star. Gas giant planets are metal-rich compared to their stars, because whether they form by core accretion or by disk instability, they gobble up rocky and icy planetesimals during the process of their formation. Ice giants and terrestrial planets are composed primarily of ice/rock and rock/iron, respectively. Stars that ate their newly formed planets would be hideous monsters that devoured their newborn, with the only evidence of the atrocities being the planets' metals left behind in the stars' atmospheres. This nightmare seemed too distressing to be an acceptable explanation for what Mother Nature might countenance.

The second idea was that the metallicity effect was the result of gas giant planets being formed by core accretion, not by disk instability. If there were more metals present in the planet-forming disk, and so more rock and ice planetesimals, then there must have been more building blocks around to assemble the roughly 10-Earth-mass solid cores that were the first step of the core accretion process. Stars that formed out of disks with higher metallicities than the typical star- and planet-forming disk might then be expected to have more gas giant planets formed by core accretion. Disk instability proceeds more or less independently of the amount of solids in the disk, because disk instability starts with a gravitational instability of the gaseous portion of the disk and so does not need to produce a strong metallicity correlation. I had published this expectation for disk instability in the *Astrophysical Journal* Letters in 2002.

The metallicity correlation was taken by many as the final proof that core accretion dominated the formation of giant planets. Little was said of the need to explain the formation of gas giants around metal-poor stars, which were known to have significant numbers of gas giants, though perhaps not so many short-period planets as the metal-rich stars. The jury's verdict was in: disk instability was out.

Pasquini and his colleagues were threatening to overturn the popular interpretation of the metallicity correlation when they submitted their paper to *Astronomy & Astrophysics.* They had decided to test the metallicity correlation by comparing the metallicities of giant stars that had been found to have extrasolar planets to those of giant stars without known planets. Surprisingly, they found that both populations of giant stars had the same range of metallicities. The metallicity correlation that applied to middle-aged stars like the Sun did not apply to older, giant stars. Metal-poor giants had just as many planets as metal-rich giants, which implied that metallicity played no role in how their planets formed.

That middle-aged stars showed a correlation with metallicity but the giants didn't was, Pasquini explained, likely to be the result of planet pollution. Giant stars are puffed up compared to middle-aged stars, with 35 times more of their mass involved in the mixing associated with their boiling. As a result, if a Sun-like star had metal-rich outer layers from having eaten planets, then by the time that star had evolved into a giant star, the iron would have been diluted by a factor of 35, and the star would no longer appear to be metal-rich. If the metallicity correlation were due to metal-rich primordial disks being better able to form gas giants by core accretion, then the entire star would be metal-rich as well, so the mixing that occurs during the giant star phase could not change the composition of the star's atmosphere. Pasquini's analysis ruled out that possibility.

According to Pasquini's results, giant stars seemed to be showing that gas giant planets could form equally well around high-metallicity and low-metallicity stars, a hallmark of disk instability. Once again, rumors of the death of disk instability appeared to have been highly exaggerated. And the fact that the rapid formation of giant planets by disk instability was expected to encourage the formation of Earth-like planets was good news for the search for living planets.

May 17–18, 2007—JPL's Navigator Program held a science forum at NASA Ames to assess the grim status of the search for habitable worlds. As a result of Griffin's drastic cuts to NASA's science budget, it had become clear that the two TPFs were likely to be replaced with a single much less effective, much less expensive "Planet Finder" capable perhaps of imaging only super-Earths, rather than Earths. Even such a mini-me Planet Finder would face a rough battle to win approval in the ongoing donnybrook over the NASA science budget.

Jon Morse, Stern's new director of SMD's Astrophysics Division, pointed out that the Navigator Program could at best hope to compete for a new mission in the next decade that would cost no more than $600 million, a sum considerably less than what was needed to finish SIM, much less get started on TPF-C and TPF-I. Morse called for a "diversified portfolio" for Navigator, a plan that would begin with relatively modest precursor missions rather than with flagship-class missions such as SIM and the TPFs. The Webb Telescope was scheduled to launch in 2013, and the next flagship mission would be a new X-ray observatory that would fly in 2020, according to Morse. At that rate, TPF might have to wait until 2028 to fly, two decades later than Dan Goldin had planned in 1998. There were even frightening rumors that NASA was quietly considering yet another (the sixth) servicing mission to Hubble (SM5) that would further extend Hubble's lease on life—at the expense of any new space telescopes.

The presentations at the science forum showed that the Navigator Program was in the process of evolving into something new. The SIM Project was trying to figure out how to build SIM for $600 million, perhaps by dropping all of SIM's science except for the planet hunt and abbreviating even the planet hunt at that. The TPF-I folks had agreed with their European colleagues on the Darwin mission that the best design for a joint infrared interferometer mission was the "Emma x-array," named after the wife of Charles Darwin. Emma

Darwin would consist of four collector telescopes with diameters of 60 inches (1.5 meters) and one device that would combine the beams from the four telescopes, for a total of five spacecraft. Emma Darwin would be able to image about 100 Earth-like worlds if every nearby star had one, at a total cost of $1.5 billion to be borne by ESA and NASA, and perhaps by Japan. Many other ideas about how $600 million could be spent to look for extrasolar planets percolated around. Everything was up for grabs again, as it was in 1988, when NASA first started thinking about planet hunting.

Malcolm Fridlund, project scientist for both Darwin and CoRoT, brought us up to date on CoRoT's status. CoRoT's cameras had demonstrated on orbit enough photometric precision (ability to accurately measure the brightness of a star) to detect transits by Earth-size planets. CoRoT was originally proposed to fly for just a few years and to spend only a few months at a time looking for transits—not long enough to detect the Earth-like planets on year-long orbits that Kepler was designed to detect. However, the CoRoT astronomers were now hoping that CoRoT would be kept operational for at least 6 years, even longer than Kepler's planned 4-year mission. Clearly, the CoRoT team was out to steal Kepler's thunder. CoRoT was a comparative bargain as well, with a total cost of about $200 million for the primary 2.5-year mission and just $10 million more for the next 3.5 years of extended mission.

Fridlund further confided the little known fact that CoRoT was carrying enough rocket fuel to operate in space for the next 20 years. There was a real chance that CoRoT would skim off the cream of new transiting planet discoveries before Kepler was even launched. Worse yet, CoRoT would still be finding new habitable worlds long after the Kepler spacecraft was shut down.

In the hunt for Earths, CoRoT was the equivalent of Sputnik I, whose launch by the Soviet Union on October 5, 1957, prompted the creation of NASA in 1958. An Extrasolar Planet Space Race was on, whether President Bush or the American public knew or cared about it or not, and Europe was ahead, just as the Soviet Union was in 1957.

May 30, 2007—The American Astronomical Society held its spring meeting in Honolulu, Hawaii, where it became clear that because of continued cost overruns at the prime contractor, Ball Aerospace, Kepler's launch date was likely to slip another four to six months beyond the planned date of November 1, 2008. Kepler team members were forced to admit that there was a good chance that CoRoT would find the first Earth-sized planet orbiting in the habitable zone of a Sun-like star. However, it was argued that CoRoT would be able to find only a few such Earths because of several factors in Kepler's favor. CoRoT was able to stare only at a patch of sky about 20 times smaller than what Kepler would be searching for transits, which would limit the number of stars that CoRoT could monitor simultaneously. The CoRoT telescope was also more than three times smaller in diameter than Kepler, so CoRoT's target stars would have to be 10 times brighter than Kepler's in order for the same photometric precision to be achieved. Both of these factors meant that CoRoT could not survey nearly as many stars as Kepler; perhaps it could survey 12,000, compared to Kepler's initial list of 170,000 or so.

June 6, 2007—Bill Borucki sent the Kepler Science Team an e-mail about a meeting the Kepler Project team had held a few days earlier with Alan Stern, Jon Morse, and NASA chief scientist (and, not coincidentally, 2006 Nobel Prize winner in physics) John Mather in Boulder, Colorado. The ongoing cost overruns by Ball Aerospace and its contractors had led to a revised total mission cost $42 million higher than the previous estimates. The Kepler Project had asked NASA headquarters for another $42 million, as was customary.

There was a big difference this time, though: Alan Stern. Stern had declared an end to mission cost overruns, and he was deadly serious about it. Stern had called for the Boulder meeting to review what was being done to replan the Kepler Mission. Even though others in attendance thought the revised plan was acceptable, Stern thought it did not reduce the cost and schedule slip sufficiently. He ordered the Kepler Project to devise a new plan and return for another review on July 6. If this second replan was not acceptable, Kepler would face the dreaded Cancellation Review, a mere formality before being led to the execution chamber.

Borucki promised the team that he would save as much as possible of Kepler's science during the second replan. We would just have to wait and see. Without Kepler, NASA would have nothing to show for over 20 years of work. Kepler had been largely built, and it would soon have been ready for the trip to Cape Canaveral, yet now it was back on life support.

June 18, 2007—A Nobel Conference was held on the subject of extrasolar planets at a plush executive retreat overlooking a fjord outside of Stockholm, Sweden. Only one or two such conferences are held each year, and the fact that this one was on extrasolar planets seemed to be a harbinger of a possible future Nobel Prize in physics for planet hunting.

The Kepler Mission science team members gathered in private at the retreat to discuss what to do to support Kepler in its latest moment of peril. The rumor was that NASA headquarters believed that the scientific return expected from Kepler was not worth the money it would cost to complete the mission, even though NASA had already spent most of the money needed to build Kepler. We quickly put together a letter to the Astrophysics Subcommittee of the NASA Advisory Council stressing that only Kepler could be certain to detect Earth-like planets, if they existed, and to determine their frequency with a high degree of confidence. Geoff Marcy fired off the letter on our behalf in an e-mail to the Astrophysics Subcommittee, and we went back to the Nobel Symposium sessions to hear the latest about CoRoT.

Malcolm Fridlund noted that CoRoT had started its first long field exposure on May 2, 2007, beginning 150 days of staring at the same 12,000 or so stars. CoRoT's camera had already stumbled on an old Delta 2 rocket stage wandering in Earth orbit. It had also seen the orbital debris from the Chinese antisatellite missile that explosively impacted a Chinese satellite on January 11, 2007, achieving success on China's fourth attempt at a kill. Fridlund pointed out that these 150 days of staring were long enough for CoRoT to detect two or three transits of planets with orbital periods as long as 60 days, and that 60-day orbital periods would place a planet in the habitable zone for K dwarf stars. Confirmation of transiting planets was to be done with a network of European telescopes in the Canary Islands, France, and Germany, and in Chile with the premier HARPS spectrograph. The CoRoT target fields were near Earth's equator so that ground-based follow-up observations could be done from observatories in both the Northern and Southern Hemispheres.

June 21, 2007—Portuguese astronomer Nuno Santos and his colleagues on the Geneva Observatory team submitted a paper to *Astronomy & Astrophysics* with the first results from their new Doppler search for planets around low-metallicity stars. Given the metallicity correlation, and the strong dependence of the two competing formation mechanisms on metallicity, understanding how the formation of gas giant planets depended on the stellar metallicity was important. The Geneva team had begun a search dedicated to low-metallicity F, G, and K dwarf stars with the HARPS spectrograph in 2003. They searched 105 stars regularly—a small sample out of the thousands of more metal-rich stars being followed, but a sample large enough to provide some fresh insight.

Santos and his colleagues discovered that the G dwarf HD 171028 had a gas giant companion with mass of at least 1.8 Jupiter masses, orbiting at 1.3 AU, a distance slightly greater than Earth's distance from the Sun. HD 171028 had about one-half of the iron present in the Sun's atmosphere, making it relatively metal-poor compared to most of the planet-bearing stars. This was only the first planet found in their metal-poor search, but the team must have had evidence for more wobbles under way, because they concluded, "The detection of an increasing number of giant planets orbiting low-metallicity stars reopens the debate about the origin of these worlds." Maybe disk instability was not such a crazy idea after all. Santos even concluded that the results on giant stars found by Luca Pasquini and colleagues suggested that planetary pollution was more important than had previously been thought to be the case, although they also raised several good objections to the idea that pollution was the *sole* explanation for the metallicity correlation. Still, low-metal stars had planets, and pollution might have helped produce the metallicity correlation, and that meant that disk instability was back in the race.

Given the large numbers of gas giants that been found already, it was odd that the battle over how they had formed was still raging with no clear outcome in sight. There was a good chance that both core accretion and disk instability were able to form gas giants, and the debate was evolving into one where the main question was which mechanism was the dominant mechanism. The initial conditions for the planet-forming disk were likely to be crucial in this regard. If the disk was massive enough, disk instability would quickly lead off with a home run or two, but if the disk was not, then it would be up to core accretion to try to bat in a few runs much later in the game.

But what about the super-Earths? Only about a dozen had been found by Doppler and microlensing, and the first transit had suggested that Gliese 436 b was a hot Neptune, not a hot super-Earth, which did not make any sense. By the time we had discovered a few hundred super-Earths, would we still be just as confused about the formation of rocky and icy planets as we were about the formation of gas giants?

June 22, 2007—Elisabeth Adams's colleagues at MIT submitted a paper to the *Astrophysical Journal* pointing out that Gliese 436 b need not be a hot Neptune consisting largely of water and water ice. Adams was able to fit the observed mass and radius of Gliese 436 b with a hypothetical planet much more similar to Earth: a planet with a core of rock and iron, surrounded by a small amount of hydrogen and helium gas in a puffy envelope.

Gliese 436 b could thus be a hot super-Earth with an outer gas giant planet sibling, such that the two resembled Earth and Jupiter—an understandable situation in terms of theories of planet formation and orbital migration. If Gliese 436 b had an atmosphere of hydrogen and helium gas, it must have managed to grow to super-Earth size while the disk gas was still there. If so, then that same disk gas could have driven Gliese 436 b to an orbit near the inner edge of the gaseous disk, where it remained. Gliese 436 b must then have formed in the inner disk, where the disk gas would have been so hot (thousands of degrees Fahrenheit) that Gliese 436 b would be able to pull on only a relatively light blanket of hot disk gas. If Gliese 436 b had formed in the cooler disk gas much farther out, it would have been in danger of pulling on so much gas that it would have become a gas giant planet.

Adams's models showing that Gliese 436 b need not be a hot Neptune made perfect sense from the point of view of the "Big Picture" of planetary system formation. If the Big Picture was right, Earths would be commonplace, some would have to be habitable, and the universe would be a much more interesting place than would otherwise be the case.

July 7, 2007—Bill Borucki sent out an e-mail to the Kepler Science Team telling us what had happened during the meeting with Alan Stern the day before. The second replan had eliminated the projected $42 million cost overrun that had stuck in Stern's craw a month earlier. NASA Ames, JPL, and Ball Aerospace had all made major cuts in the funds they expected to receive in support of Kepler. The most challenging preflight testing of Kepler was dropped in favor of testing on a smaller scale, people were laid off, and the length of the observations in orbit was cut from 4 to 3.5 years, among other reductions. Stern accepted the replan but warned against any further cost overruns. Kepler had dodged the bullet.

Even with the second replan, Kepler's total cost of well over $500 million was way out of line with the cost cap for principal investigator–led Discovery-class missions, so Stern declared Kepler to be the first of a new class of NASA missions, "Strategic Missions," and removed Borucki as the principal investigator for Kepler. Instead, Borucki would be the science principal investigator, and the JPL project manager, Leslie Livesay, would have the overall budget and management authority for Kepler.

Borucki was forced to sacrifice himself as principal investigator to keep Kepler alive, but the Kepler Science Team unanimously agreed to proceed as though Bill were still the principal investigator of his Discovery Mission, as he deserved to be.

July 25, 2007—On United Airlines flight 947 from Washington Dulles International Airport to Los Angeles International, I overheard a voice talking to other passengers about NASA's plans for the future. The voice sounded strangely familiar, so I turned around and peered through the cracks between the seats. I was surprised to see former NASA Administrator Dan Goldin telling the other passengers that he did not particularly care about the fate of SIM. Goldin was concerned instead about the fate of the Terrestrial Planet Finders, by now named TPF-1 (and, it was anticipated, TPF-2) instead of TPF-C and TPF-I. In Goldin's view, the main value of SIM was to find targets for the TPFs. If necessary, TPF-1 could search the nearby stars for its own Earths. Goldin did not mention Kepler, the only space telescope that NASA had any hope of launching any time soon.

August 3, 2007—Bill Borucki held a telephone conference with the Kepler Science Team to tell us more about that fateful meeting with Alan Stern. Kepler's total cost had risen to $572 million and could rise no more, at least not if Stern was still running SMD. The launch date had slipped by several more months, to February 16, 2009.

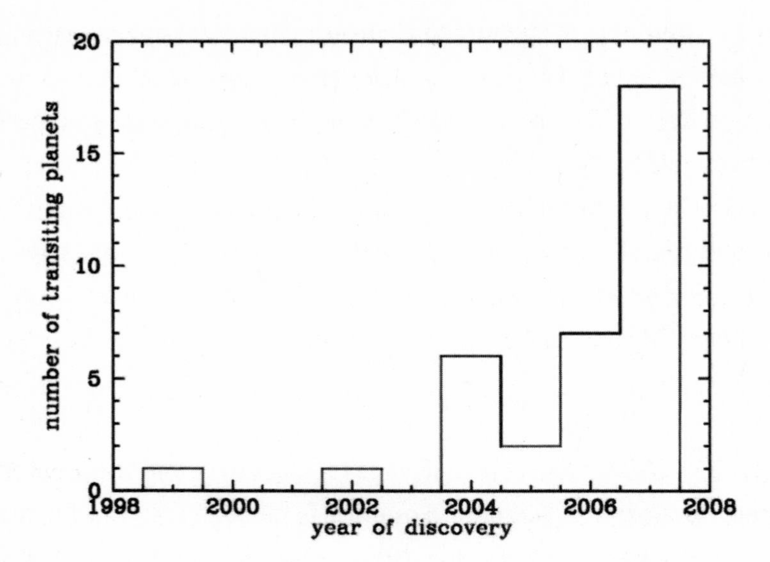

FIGURE 31. The number of transiting planets detected each year underwent a major increase in 2007.

Kepler was now planned to operate for only 3.5 years in space, barely long enough to catch the four transits by Earths. However, there was always the hope that if Kepler worked as well on orbit as planned, an extension to the nomimal mission lifetime would be granted. This would take the approval of a NASA Senior Review Board after Kepler had been in orbit for several years. In anticipation of a successful Senior Review, Kepler's fuel tanks would be launched containing enough fuel to support a 6-year mission that would nail down the frequency of habitable worlds once and for all.

October 24, 2007—Alain Leger of the University of Paris South in Orsay, France, sent out a worldwide e-mail bemoaning the decision by ESA not to select Darwin for flight during the time period 2015–

2025. Leger was an active supporter of the Darwin mission concept in Europe, and the decision to defer Darwin meant that its launch would slip by at least another 10 years. ESA's "Cosmic Vision" for 2015–2025 did not include new worlds, although ESA did recommend that the technology needed to build Darwin should continue to be supported. ESA had cited the lack of technological readiness as the key reason for its decision.

Emma Darwin would now have to hope for a launch in 2025–2035, a time frame that seemed impossibly distant, at least for someone of my age who had been involved in NASA's planet search efforts since 1988. Leger remained hopeful and called for continued work. But it was becoming evident that Emma Darwin would need young astronomers and engineers to be involved in her development if they were to have a good chance of living long enough to enjoy the discoveries that Emma would surely make.

October 31, 2007—Jean Schneider of the Paris Observatory, who maintains an unofficial but very oft-cited web site listing all the extrasolar planet discoveries, sent out an e-mail to his exoplanet exploder list announcing the discovery of three new transiting planets (WASP-3 b, WASP-4 b, and WASP-5 b), a remarkable feat.

After a somewhat slow start, ground-based transit detection groups were cranking out the discoveries at an accelerating rate. There were numerous transit searches under way around the world, because only modest-sized telescopes were required and because precise photometry is easier to perform than the precise spectroscopy required by the Doppler technique. Confirmation of transit candidates still required high-precision Doppler spectroscopy, though, to rule out the false alarms caused by binary star interlopers and other intruders.

The three new WASP planets were found by a team centered at the University of St Andrews in Scotland. Their Wide Angle Search for Planets (WASP) program began in 2000 with a commercial 10-inch Meade telescope on the Canary Islands. Andrew Collier-Cameron of St Andrews and the WASP team had published their first two discoveries, WASP-1 b and WASP-2 b, earlier in 2007, and now they were up to five hot Jupiters. More were in the pipeline, waiting for confirmation. Other teams had done equally well. Tim Brown's Trans-Atlantic Exoplanet Survey (TrES) team had found four hot Jupiters by October 2007. Gaspar Bakos of the Harvard-Smithsonian Center for Astrophysics had started the Hungarian-made Automated Telescope (HAT) transit search in 2003, using a collection of six small telescopes in Arizona and Hawaii. He and his team had found six transiting hot Jupiters by 2007, named, appropriately enough, HAT-P-1 b through HAT-P-6 b.

Transits had become the name of the game, in large part because they offered the opportunity to determine a planet's true mass and radius, and therefore its density, as well as the chance to learn about the planet's atmosphere during the primary and secondary eclipses. This meant that we knew a lot more about the several dozen transiting planets than about the hundreds of Doppler-only planets. We knew even less about the handful of mysterious microlensing planets and their host stars, which typically bent and brightened the light from a background star for periods ranging from a few days to a few months but then disappeared from sight, never to repeat this feat again. The advice being whispered to young astronomers at beer hour was a single word: not "plastics" but "transits."

CHAPTER 9

300 Solar Systems and Counting

... NASA is not solely, or even primarily, about science.

—MICHAEL GRIFFIN (2008)

November 7, 2007—The Kepler Science Team held a meeting at Ball Aerospace in Boulder, allowing us to see the primary mirror and various other key telescope components ready to be assembled into the Kepler Space Telescope. Funding for the Science Team had been cut along with everything else during the most recent replan, but the Kepler Science Team was steadfast in their support for the mission: Kepler was the only game in town for astronomers interested in finding Earths with a space telescope that was in some danger of being launched during their all-too-finite lifetimes.

Borucki noted that even with the latest replan, Kepler would still be able to determine the frequency of Earth-like planets, as long as this frequency was greater than about 5%. That is, if Earths were as common as we believed on the basis of the last 10 years of extrasolar planet discoveries, Kepler would determine their frequency. If Earths were less common than 5%, Kepler might find nothing at all. If Kepler found nothing Earth-like, we would not know whether Earths were truly rare or whether they simply occurred perhaps 1% of the time and Kepler just hadn't been lucky enough to catch one. We would not know how to proceed with the next step to finding and learning more about other Earths if Kepler produced a null result. A null result might very well halt SIM and the TPFs for decades to come.

Jason Rowe, a postdoctoral fellow for the Kepler Mission at NASA Ames, updated us on the progress of MOST, the Canadian 6-inch (15-cm) space telescope launched in 2003. MOST had studied some 900 stars over the last 4 years, and it was still working well, although the small eight-person science team had discovered that MOST had a problem with stray starlight. Apparently the satellite housing had cracked during the launch by the ex-Soviet ICBM. Rowe and the MOST Science Team had used the diminutive space telescope to observe transits by the first transiting planet, HD 209458 b, hoping to detect starlight reflected off the surface of HD 209458 b just before and after secondary eclipses. MOST tried and tried to see this visible light, but in the end all MOST could do was set an upper limit on the amount of light reflected from the planet.

MOST was expected to last a few more years in orbit before the inevitable damage by cosmic rays would silence Canada's version of Hubble. In the meantime, Canadian citizens were invited to suggest new targets for MOST to observe, offering them a chance to work with Canada's first space telescope, with MOST now standing for My Own Space Telescope.

December 13–14, 2007—The SIM Science Team held a meeting at the U.S. Naval Observatory in Washington, D.C., an appropriate venue for a meeting about an astrometric space telescope, given that

Naval Observatory astronomers have been making precise astrometric measurements for use in navigation for over a hundred years. Although most of the serious astronomical work was moved to a dark site near Flagstaff, Arizona, in 1955, the 26-inch refractor that was used in 1877 by Asaph Hall to discover the two Martian moons, Phobos and Deimos, is still in operating condition. During the decade from 1873 to 1883, the Naval Observatory's 26-inch Great Equatorial Telescope was the world's largest telescope built with lenses, rather than mirrors, to focus the starlight.

Visitors who attend the Naval Observatory Monday night star parties are allowed to tour the dome with the 26-refractor, but only if it is cloudy. On clear nights, the 26-inch is still used to study the orbits of binary stars, as it has been ever since 1873. While still in office, Vice President Al Gore liked to wander over at night to see how the astronomers were doing; the vice president's residence is located in the former Naval Observatory Superintendent's house, a short walk from the dome. Vice President Dick Cheney did not wander over. During Cheney's tenure, the grounds were turned into a fortress with significantly reduced access. A considerable number of fully loaded dump trucks were observed departing the USNO grounds, unequivocal evidence of the bunker mentality of the second Bush administration and of the creation, deep beneath the vice president's residence, of a fortified retreat.

An astronomer who had been working in the dome happened by chance upon Vice President Cheney and his escorts late one night, and he saluted the vice president with his middle finger. Cheney ignored the salute, but the astronomer was banished to Flagstaff for his impertinence.

In keeping with the desperate state of affairs created by the Bush administation, the main focus of the SIM Science Team meeting was what to do to keep SIM alive. The Navigator Program Science Forum in May had made it clear that NASA headquarters was not planning on flying SIM any time soon, if ever. The best that Jon Morse had been able to offer the planet-hunting community was a shot at a $600 million mission in the next decade. The cost estimate for completing SIM was $1.9 billion, in addition to the $500 million already spent on developing the unprecedented technology needed to make SIM work as originally envisioned by Michael Shao—that is, to find Earth-like planets.

As a result, the SIM Project at JPL was forced to think about what they could do for a measly $600 million more. JPL came up with the idea of a stripped-down version of SIM, called SIM Planet Hunter (SIM-PH), which would use the same technology as the full-up version of SIM but not as much of it. The intent was, by cutting this and that and shrinking everything, to shoehorn SIM-PH into the $600 million box. SIM-PH would do nothing but hunt for Earths around the best (usually the closest) 65 stars and look for more massive planets around 1000 or so others. The other astronomical goals that drove SIM's creation would have to be sacrificed on the altar of NASA's new-found frugality.

To many, it looked like the full-up version of SIM had just been descoped into SIM-PH, a move that was certain to enrage the astronomers on the SIM Science Team and in the general community who were counting on SIM to make amazing advances in their areas of interest, the most obvious being the cosmic distance scale. SIM's astrometric agility enables it not only to find Earth-mass planets, but also to measure the distances of celestial objects with unmatched precision. With SIM's measurements of parallaxes, astronomers would be able to determine the distance to any bright star in the Milky Way galaxy to an accuracy of about 10%. SIM would be able to obtain accurate distances to stars that are 1000 times farther away than the farthest measured by the previous astrometric record holder, the European Hipparcos space telescope launched in 1989. SIM could thus hope to extend and improve the cosmic distance scale, with implications for the expansion rate and ultimate fate of the universe. SIM-PH would not.

JPL's director, Charles Elachi, had said that JPL was still committed to the full-up version of SIM and hoped to win support for SIM from the upcoming NAS Decadal Survey of astronomy. SIM had first been blessed by the 1991 Decadal Survey, and the 2000 Survey had assumed that SIM was on its way to being launched, but now SIM would have to be submitted to the 2010 Survey process slated to begin in late 2008. After having successfully navigated the previous two Decadal Surveys, SIM would be subjected to a new trial by fire.

JPL's SIM project manager, Jim Marr, presented three options at the USNO meeting, all of which managed to meet the minimum requirements set for SIM by the previous NAS Decadal Surveys, in spite of vastly different costs. The original SIM concept had been purged of many of its bells and whistles during a NASA headquarters-ordered replan a year earlier and had shrunk in size from 33 feet (10 meters) to 30 feet (9 meters), but it was still estimated to require another $1.9 billion. The planets-only version, SIM-PH, was a bargain in comparison for only $700 million, but it was still well over Jon Morse's $600 million figure. The intermediate option was dubbed SIM-Lite, using the spelling that appeals to thirsty middle-aged beer drinkers with expanding waistlines. SIM-Lite would cost about $1 billion, considerably less than full-up SIM, as a result of having been shrunk by a further 33%: SIM-Lite would be equivalent to a space telescope with a diameter of 20 feet (6 meters) instead of 30 feet (9 meters). SIM-Lite was compact enough to fit into a smaller launch vehicle, and that saved money too. SIM-Lite was SIM cut down to the bare bones of what was needed to meet the NAS Decadal Survey requirements for astrometric accuracy.

SIM-Lite could still do much of the science planned for SIM, although it would take SIM-Lite appreciably longer to achieve the same accuracy as the full-up SIM, and this in itself limited the number of targets that could be studied. SIM-Lite would spend half of its time looking for planets and the other half looking at stars all across the Galaxy. If Earth-like planets were orbiting every nearby star, SIM-Lite could expect to find about 60 Earth-mass planets. If Earths were 10 times less common, SIM-Lite would find 6 Earths, and if Earths were 10 times less common still, SIM-Lite might not find any Earths at all. Kepler's task was to determine just what the frequency of Earths was, so that critical decisions about expensive space telescopes such as SIM and the TPFs could be made in a rational manner. It was no wonder that Alan Stern had reclassified Kepler as the first Strategic Mission for NASA. NASA headquarters needed to know the answer to the question that only Kepler could provide.

Michael Shao's creation, SIM, was fighting to stay alive. Shao's hair had greyed since I first encountered him at the Planetary Systems Science Working Group meetings in 1988, but he retained the same fervor for bringing SIM through this prolonged gestation period and into the world. Shao could only sit and listen to the debate, his knees jiggling up and down, and answer questions when obscure points arose that only he could address. Junior members of the JPL SIM team tended to glance periodically at Shao as they gave their presentations, seeking approval from SIM's ultimate guru.

One of the dangers facing SIM was a competing space astrometry mission, called GAIA, that the Europeans were developing and planning to fly in late 2012 for a 5-year mission. GAIA was originally named the Global Astrometric Interferometer for Astrophysics, but the ESA astronomers had decided that space interferometry was still too much of a challenge and therefore adopted a non-interferometric design for the space telescope. The acronym remained, although GAIA is often spelled as Gaia by ESA, which makes the mission appear to be named after the Greek goddess of the earth. GAIA was intended to follow the Hipparcos mission, which surveyed a million stars. GAIA would outdo Hipparcos a thousand times by surveying a billion stars, roughly one out of every hundred stars in our galaxy.

GAIA was a formidable opponent for SIM, but the two missions were designed to accomplish different tasks. Whereas GAIA was planned to create a catalog of the locations and approximate distances of a billion stars, SIM was a "pointed mission" that studied a much more limited number of target stars, perhaps only 10,000 or so, but looked at those target stars long enough and often enough to determine their distances to extraordinary accuracy. SIM would also be able to detect Earth-mass planets, whereas GAIA would be able to find only Jupiters.

The SIM Science Team agreed to support the SIM-Lite concept as their preferred option, and Caltech's Shri Kulkarni, acting as the Science Team leader, made the case for SIM-Lite to Jon Morse when Morse arrived on the second day of the meeting. Kulkarni emphasized that SIM needed to team up with GAIA in order to maximize the scientific productivity of both missions: GAIA would do the overall survey of the Milky Way galaxy and find interesting targets for which SIM could then perform more detailed measurements.

Jon Morse listened intently, but in the end he could not promise that SIM would survive the next year. With a flat or declining budget in the Astrophysics Division of SMD, Morse said that until the Webb Telescope was launched, it was difficult for NASA to keep expensive future missions on life support indefinitely. The implications for SIM were ominous.

December 17, 2007—The web site SPACE.com managed to obtain a copy of the omnibus appropriations bill for NASA for FY 2007–08. The budget report was the product of the joint House-Senate Appropriations Committee charged with finding common ground between the budgets passed independently by the two houses of Congress.

Like many government agencies, NASA had been operating on a "Continuing Resolution" since the new fiscal year began on October 1, as a consequence of the perennial inability of Congress to enact budget legislation on time. Continuing resolutions basically mean "Do not spend any more money than you received last year, and maybe even less, so that you will not get caught short if Congress should decrease your agency's budget." Given that 2008 was an election year, Congress was trying to improve its sorry performance and get the budget finished earlier than usual, though still several months late.

The news for NASA was dreadful. The Senate bill had specified an extra $1billion in emergency funding to help NASA pay the costs associated with the return to flight after the Columbia shuttle disaster, thanks to the efforts of Senators Mikulski and Kay Bailey Hutchison of Texas, home of NASA's human space flight effort. These emergency funds were dropped from the final conference report. The final mark for NASA was $17.3 billion, seemingly a large figure, but less than had been requested in either the House or the Senate bill, even without the emergency $1 billion. For comparison, the Bush administration's occupation of Iraq was estimated to be costing the United States about $12 billion every month. Perhaps TPF would have had a higher priority in the Bush administration if its acronym had stood for Terrestrial Petroleum Finder.

NASA's Science Mission Directorate received $5.6 billion, a cut of about $100 million from the two House and Senate bills. The report language chided the administration for asking for only a 1% increase in NASA spending for science for the current and future fiscal years but then punished NASA science by allocating SMD even less funds than the House or Senate had proposed. The Appropriations Committee report warned NASA not to ignore the report's language and then proceeded to set floors for the spending on several specific NASA missions. Hubble was to receive no less than $280 million in FY2008, whereas Webb would receive at least $545 million. Maryland would not suffer too badly, as long as Senator Mikulski was still around. The final report also called for a floor of $626 million for the Mars Exploration Program. The last item on Congress's wish list was $60 million for SIM.

$60 million for SIM? The joint report noted that the $60 million was an increase of $38 million over what had been requested for SIM in the president's budget. The report further noted that SIM had passed muster with the NAS Decadal Surveys in 1991 and 2000, had completed all of its technology milestones, and was ready to fly. The congressional report explicitly directed NASA to stop fooling around and begin the development phase of SIM.

December 20, 2007—Jean Schneider sent out an e-mail five days before Christmas announcing the second transiting planet discovered by CoRoT. CoRoT-Exo-2 b had a mass of 3.5 Jupiter masses and orbited its solar-mass host star every 1.7 days, making it another hot Jupiter. However, its radius was only 1.4 Jupiter radii, meaning that

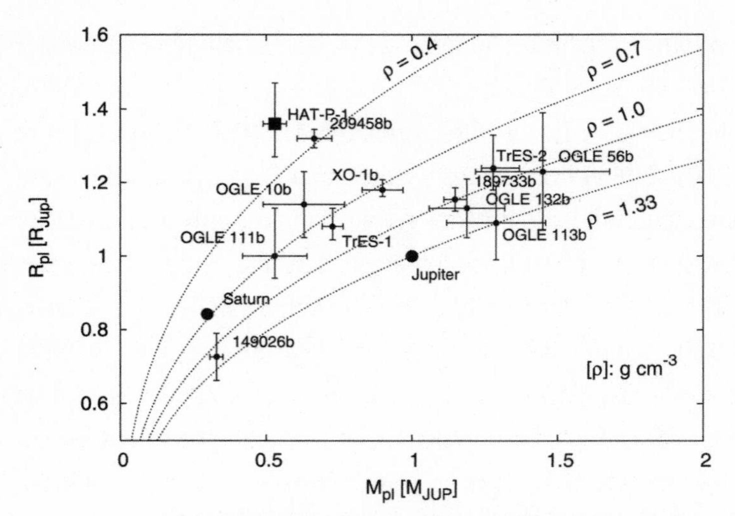

FIGURE 32. The sizes (radii in units of Jupiter's radius) and masses (in units of Jupiter's mass) of the transiting planets show that hot Jupiters have a wide range of densities. [Reprinted, by permission, from G. Bakos et al., 2007, *ApJ*, volume 656, page 557. Copyright 2007 by the American Astronomical Society.]

its density was considerably greater than that of CoRoT-Exo-1 b, which had one-third the mass of CoRoT-Exo-2 b but an even larger radius. Whereas CoRoT-Exo-1 b was a fluffy hot Jupiter, CoRoT-Exo-2 b was not. As usual, Mother Nature was keeping us on our toes by revealing the diversity of extrasolar planets, one planet at a time. What would CoRoT find next?

January 4, 2008—*Science* magazine rated the chances of six top federal science officials surviving the transition to the new administation that would arrive on January 20, 2009. Bush's one and only science advisor, John Marburger, was given the lowest odds of all: 0%. Arden Bement, the popular head of the National Science Foundation, was given the best odds of surviving: 50%. Michael Griffin did not fare nearly as well, garnering odds of only 20%, given his promise to replace the shuttle with a new vehicle while still maintaining the integrity of the rest of NASA, an impossible task without a major infusion of funds.

January 8, 2008—The American Astronomical Society held its winter meeting in Austin, Texas, and as usual invited Michael Griffin to deliver an address about the state of NASA science. Griffin started off by complimenting astronomers for what they had accomplished in seeking to answer the eternal questions: how the universe began, how the Solar System formed, whether other planetary systems existed, and whether life might exist on them. He singled out the example of the Hubble Ultra Deep Field image, which revealed thousands of young galaxies, and pointed out that he had worked as an engineer on Hubble's fine guidance system back in 1983.

Griffin noted that although NASA had been spending on average $1.5 billion each year on astrophysics since 1999, the plan for the next 5 years was to spend $1.2 billion per year, in equivalent dollars. This was a cut of $300 million per year, but Griffin argued that it was still a lot of money compared to what the National Science Foundation, for example, was spending on astronomy. He pointed out that other areas in NASA had suffered as well, particularly the human space flight program, which was facing a 5-year hiatus between the last shuttle flight in 2010 and the first flight of its replacement vehicle to the International Space Station in 2015. In between, the United States might have to depend on Russian rockets for rides to the Station. Griffin even likened U.S. astronomers to children for preferring their own narrow scientific interests to an American space policy focused on the Space Station, courtesy of President Bush's Vision for Space Exploration. Griffin said that although science is a "very important part of NASA," "NASA is not solely, or even primarily, about science." In other words, sit down and shut up, kids.

In between these brutally frank remarks, Griffin addressed one of the spending floors in the omnibus appropriations bill: the one for SIM. Griffin acknowledged that he would like to proceed with flying SIM (he too wanted to know whether we are alone in the universe) and pointed out that Kepler would be launched in early 2009. However, NASA simply could not afford SIM in an era when Hubble, Webb, SOFIA, and research grants consumed 85% of SMD's budget. If SIM were started in 2008, as directed by Congress, the funds would have to come from elsewhere in SMD. Griffin went with the Department of Interior's classic Washington Monument ploy. When faced with budget cuts, Interior sadly announces that it will have to shut down the Washington Monument until further notice—that is, until their budget requests are met. Griffin went with NASA's Washington Monument: the Webb Telescope. If SIM went forward, then Webb, or perhaps some other NASA mission, would be delayed.

Griffin noted that Congress did not come up with the SIM demand on its own; rather, "external advocacy" had been involved. He urged astronomers to avoid such advocacy in the future and to rely on the NAS Decadal Surveys for prioritizing space missions. Griffin asked the upcoming Survey to rank all missions that had not yet entered development, even if they had been previously approved (exactly the position SIM was in) and to include cost caps in their rankings. "There is no free launch," Griffin said in a brief attempt at levity in what was otherwise a disturbing speech.

January 9, 2008—The SIM Project held a special evening session at the Austin meeting about what SIM Lite could accomplish for astronomy in general and planet hunting in particular. The session was packed with astronomers who had been spurred to attend in part by Griffin's Washington Monument ploy the previous day. After the formal presentations, the audience had their say. Matt Mountain, director of the Space Telescope Science Institute, with responsibility for both Hubble and Webb, bluntly attacked SIM for a lack of community involvement. Others complained that SIM had started out as an intermediate-class mission but had grown in cost to a flagship-class mission in the interim. The much greater cost growth of Webb did not seem to matter.

SIM would be facing a hostile audience at the upcoming Decadal Survey. The main question was likely to be who would be chosen to sit on the key Decadal Survey committees. A hostile audience is an annoyance, but a trio of hostile ring judges could end SIM's boxing career just as well as a knockout punch.

January 15, 2008—*Nature* announced online the source of the omnibus appropriation language in support of SIM: Congressman Adam Schiff, representing Pasadena, California. Schiff was put on the House Appropriations Subcommittee responsible for NASA in 2007, and he wasted no time in successfully representing his Pasadena constituents (among them Caltech, which manages JPL for NASA) in the 2008 budget brouhaha. Schiff appeared to be doing for Southern California what Senator Mikulski had for decades done for Maryland: bringing home the bacon-flavored tofu.

February 7, 2008—Richard Kerr of *Science* magazine called me to find out what I thought about a microlensing planet discovery to be

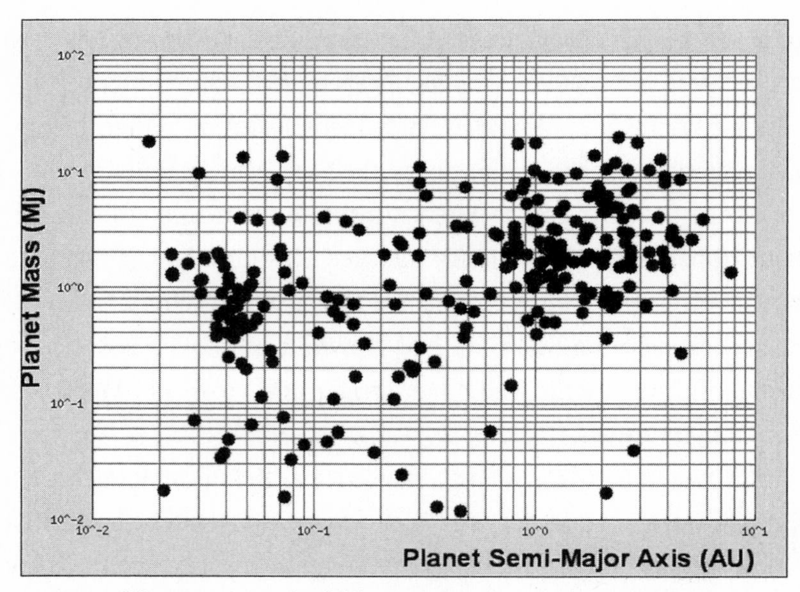

FIGURE 33. Discovery space for 266 extrasolar planets, showing their masses (minimum masses for the planets discovered by Doppler wobbles that do not transit) and orbital distances (1 AU is the Earth–Sun distance), both axes plotted on logarithmic scales to allow the large range in masses and distances to be discerned. [Courtesy of Jean Schneider, Paris Observatory].

published in the February 15 issue. Dick sent me an embargoed copy of the paper, but I already knew about the discovery, because David Bennett had mentioned the result during his talk in the session on exoplanets I had chaired the previous summer at the Gordon Conference on the Origins of Solar Systems, held at Mount Holyoke College in Massachusetts. Gordon Conference talks are supposed to be off the record, so Bennett's remarkable claim could not be discussed elsewhere until he was ready to make the work public. The time had come.

The *Science* paper was authored by B. Scott Gaudi of Ohio State University, Bennett, and 68 other named astronomers, mostly members of five different microlensing collaborations (OGLE, MicroFUN,

MOA, PLANET, and RoboNET), as well as a few stragglers who tagged along for the ride. Gaudi, Bennett, and their colleagues had found something remarkable: the first double planet system discovered by microlensing.

Not only was it the first double planet system found by this technique, but the system appeared to be weirdly analogous to the two gas giants in our own Solar System. The microlensing event OGLE-2006-BLG-109 had begun on March 28, 2006, with an e-mail from the OGLE team to the MicroFUN and RoboNet groups asking them to watch this event, because it was likely to involve a planet. Sure enough, on April 5 and 8, brightenings indicative of a planetary companion were observed and attributed to a Jupiter-mass planet. The real surprise was that in between these two brightenings of the lensed background star, another brightening occurred on April 5–6 that could have been caused only by a second planet, again of Jupiter-mass. This amazing discovery was made possible through around-the-clock observations of OGLE-2006-BLG-109L by the five worldwide microlensing networks working in tandem.

At the time of the Gordon Conference talk, Bennett and his team had estimated that OGLE-2006-BLG-109Lb was a planet with a mass of 0.99 Jupiter masses orbiting 2.7 AU from its host star and that OGLE-2006-BLG-109Lc was a 0.36-Jupiter mass planet orbiting at 5.5 AU. By the time the *Science* paper was published, their best mass estimates had been lowered slightly, to 0.71 and 0.27 Jupiter masses, respectively, and the orbits had moved inward somewhat, to 2.3 and 4.6 AU. The host star was estimated to have a mass about one-half that of the sun; it was probably an M dwarf star. The *Science* paper made the point that the OGLE-2006-BLG-109L system looked just like a scaled-down version of the Jupiter-Saturn system in that the mass ratios and orbital distance ratios of the two exoplanets (2.6

and 0.50, respectively) were close to those of Jupiter and Saturn (3.3 and 0.55, respectively). Given that these were just the fifth and sixth microlensing planets detected to date, it was clear that M dwarf stars with one Jupiter-mass planet had a pretty good chance of having another Jupiter-mass planet. Solar system analogues might then be expected to be commonplace around M dwarfs, the most numerous type of star in the Galaxy. The universe had just become a little more crowded.

The *Science* paper stated that the new results were consistent with the formation of these two gas giant planets by core accretion, in spite of the fact that the 2004 *Astrophysical Journal* paper by Laughlin, Bodenheimer, and Adams had pointed out the difficulty that attended the formation by core accretion of gas giants around increasingly lower-mass stars, such as the one in the OGLE-2006-BLG-109L system. Beaulieu and Gould's account of their 2006 microlensing discovery of the two cold super-Earths had cited the Laughlin paper as evidence for the failure of core accretion to produce gas giants around M dwarfs, but the microlensing folks seemed to have forgotten about Laughlin, Bodenheimer, and Adams in the intervening two years. Instead, the *Science* paper quoted a 2005 *Astrophysical Journal* paper on core accretion by Shigeru Ida of the Tokyo Institute of Technology and Douglas Lin as support for the OGLE-2006-BLG-109L system. However, the Ida and Lin paper had confirmed the results of Laughlin, Bodenheimer, and Adams, making the OGLE-2006-BLG-109L gas giants system problematic for formation by core accretion.

So how did the two gas giants form in the OGLE-2006-BLG-109L system? If core accretion was not up to the task, that left only disk instability, the perpetual understudy, waiting offstage as usual in case the star should fall and break a leg. Disk instability has no particular problem making gas giants around lower-mass stars, because it is such a rapid process compared to core accretion, provided that the lower-mass star has a disk massive enough to become gravitationally unstable. Best of all, disk instability predicted that a system that formed in the way the Solar System may have formed, in a region of high-mass star formation under a withering spotlight of ultraviolet radiation, should have a Jupiter-Saturn pair with orbital distances that scaled with the mass of the central star. OGLE-2006-BLG-109L had about half the mass of the Sun, so the critical distance for photoevaporation should be roughly half that distance in our Solar System. In our Solar System, that distance is Saturn's orbit. In the OGLE-2006-BLG-109L system, the Saturn-mass exoplanet orbited at half of Saturn's distance. The new system was thus fully consistent with the disk instability scenario. However, the *Science* paper made no mention of this alternative explanation, even though it had been published in two *Astrophysical Journal* papers in 2006. Disk instability was still largely ignored in planet formation theory, struggling like Rodney Dangerfield for a little respect.

February 8, 2008—The president's budget request for NASA for the next fiscal year, 2009, increased the funds for NASA's science programs by only 1%, less than the annual rate of inflation. NASA's science funding was entering a period of slow but steady decline.

February 25, 2008—Another SIM Science Team meeting was held at the U.S. Naval Observatory in Washington, D.C., where it became clearer how NASA headquarters had handled Congress's demand that NASA headquarters spend at least $60 million for SIM in the

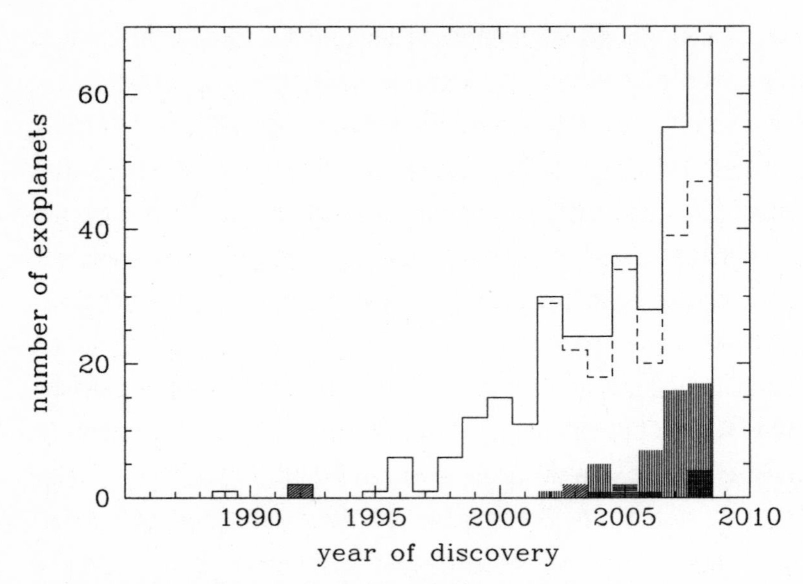

FIGURE 34. **Total number of exoplanets discovered each year (solid line), as well as the total discovered each year by Doppler wobble (dashed line), transits (vertical stripes), microlensing (horizontal stripes), and pulsar timing (oblique stripes).**

current fiscal year. NASA headquarters had countered with an offer to spend $24 million on SIM, as well as $8 million in funds carried over from the previous year, for a total of $32 million, not $60 million, but significantly more than the $22 million that NASA had originally been requesting for SIM. The rest of the $60 million disappeared somewhere inside SMD, although some would go to other planet search efforts. Congress agreed, and that was that. SIM would survive for another year, but without the large boost in funds needed to start development.

The SIM Project realized that the full-up version of SIM, with a cost of $1.9 billion, was a non-starter. Even SIM-Lite at $1 billion might cause the Decadal Survey to rank SIM low enough that it would never fly. SIM would be competing for a slot as a Strategic Mission—the new class of missions inadvertently created by Kepler's cost overruns, missions costing more than $600 million but less than a multibillion-dollar flagship mission like Webb. The Decadal Survey deliberations would start in the fall of 2008, so the SIM team had a few months to refine and polish their concept of SIM and to try to find the sweet spot where SIM returned the most science per dollar expended. Any other outcome was likely to bode ill for SIM.

After two decades of planning, studying, and technology development, NASA's hopes for flying a dedicated space telescope capable of studying Earth-like planets either by astrometry (SIM) or by imaging (TPF-C, TPF-I) was back to square one, subject to approval and ranking by yet another NAS Decadal Survey committee. In spite of transit searches having been dismissed in the 1990 NAS report on detecting extrasolar planets, the Kepler Mission was NASA's one and only registered entry in the race with Europe to discover new Earths.

March 20, 2008—Water was back. *Nature* published a letter by Mark Swain and Gautam Vasisht of JPL and Giovanna Tinetti, now of University College London, asserting a new claim for the presence of water in the atmosphere of a hot Jupiter. The year before, Swain had been the leader of one of the three teams that looked for but failed to see any evidence for water in the spectra of two hot Jupiters, HD 209458 b and HD 189733 b. The three previous searches had used Spitzer to measure the infrared light given off by the hot Jupiters just before and just after they passed behind their stars, during secondary eclipse, but this time the searchers used Hubble to look during the primary eclipse, when some of the star's light would pass through the planet's upper atmosphere on its way to Earth. Tinetti and her colleagues had used the same technique to make a claim for water in the atmosphere of HD 189733 b the previous year as well, but those measurements were made with Spitzer. Tinetti's claim had been called into question by David Ehrenreich and some of the same authors who were on Tinetti's paper, making the claim for water on HD 189733 b muddy at best. Swain and Tinetti decided to try their luck with Hubble this time on HD 189733 b.

The key instrument for their observations was the Near Infrared Camera and Multi-Object Spectrometer on Hubble. Although the camera was limited to shorter infrared wavelengths than Spitzer, the near-infrared spectrum of a hot Jupiter was predicted to show the presence of absorption of the star's light around 1.9 microns, caused by water vapor in the planet's atmosphere. The spectrum of HD 189733 b during its primary eclipse extended from 1.5 to 2.5 microns and showed clearly the expected 1.9-micron absorption feature. HD 189733 b had water after all.

And that wasn't all. In order to explain the bumps and dips in the NICMOS spectrum, the team found that they had to add the absorption caused by a second molecule, methane. Methane is one of the four primary biomarkers (water, carbon dioxide, oxygen, and methane) for an inhabited world. Swain, Vasisht, and Tinetti had won the quinella bet of the extrasolar planets sweepstakes by finding both molecules in the atmosphere of the same hot Jupiter. The detection of methane and water on HD 189733 b was another promising step in the direction of detecting signs of life on another world.

March 24, 2008—As a result of major and ongoing cost overruns in NASA's Mars Exploration Program, principally for the Mars Science Laboratory, whose cost had grown from $750 million to $2 billion,

Alan Stern decided that the Mars Program would have to suffer some pain. A letter was sent to the Mars Exploration Program office at JPL telling it to cut the budget for the Mars Exploration Rovers (MER), two robots that had been exploring the Martian landscape since January 2004. The Mars Rovers Mission was to be cut by $4 million in the current fiscal year, a 40% cut in the funds remaining in the fiscal year. The Mars Rovers team responded with the usual Washington Monument ploy. Cornell's Steven Squyres, the MER principal investigator, stated that with a cut of that size, they would have to turn off one of the rovers, at least temporarily. Things might get even worse in the next fiscal year.

SPACE.com reported the cuts with the eye-catching headline "Budget Cuts Could Shut Down Mars Rover," something that would bring pain to nearly every sentient being who had been following the two wandering rovers for the last several years. The rovers had been faithfully sniffing around Mars, looking for signs of water, and now at least one was being threatened with being put down.

March 25, 2008—NASA headquarters rescinded Stern's order to cut the funds for the Mars Rovers program. Michael Griffin personally intervened. Griffin noted that his office, which normally does not get involved at the level of individual ongoing missions, was not informed of SMD's plan to cut the funds. He decided that both rovers were to be kept alive, which was bound to please the public. After the debacle over Sean O'Keefe's decision to drop the last servicing mission to Hubble, it was clear that Griffin wasn't about to open that third envelope any earlier than he had to. He had already reorganized NASA headquarters, in part by bringing in Stern to head SMD, so only the third envelope remained in his possession.

March 26, 2008—Alan Stern announced his resignation as the head of NASA's Science Mission Directorate. Griffin had tried to get Stern to change his mind, but Stern did not appreciate having his decisions overturned from on high. Chief scientist John Mather decided to bow out as well and return to Goddard Space Flight Center to lead the development of Webb.

Without missing a beat, Griffin announced that Ed Weiler would leave his position as director of Goddard and return to NASA headquarters to run SMD, a directorate somewhat expanded from the one Weiler had been demoted from by O'Keefe in 2004. Weiler was one of the architects of NASA's plans for searching for habitable worlds, from Kepler to SIM to the two TPFs. Planet hunters could not ask for a better head of SMD than Ed Weiler. Weiler's appointment was only an interim appointment, but perhaps "interim" would last long enough to get planet hunting beyond Kepler back on NASA's priority list. Given the SMD budget problems that Weiler would inherit from Stern, Griffin, and the Bush Vision for Space Exploration, though, it was hard to be optimistic that much could be done to revitalize looking for Earths.

April 1, 2008 (April Fool's Day) —The nasawatch.com web site posted a link to "Today's Video," showing Steve Squyres in a lively appearance on *The Colbert Report* as if it had just occurred the night before, even though the episode dated from June 7, 2006. Stephen Colbert had been relatively merciful with Squyres, actually letting him speak at times and highlighting a model of the Mars rovers sitting on the table between them. Griffin's instincts had been right: if Comedy Central liked the Mars rovers, it was best not to touch the third rail on their track.

On the same day, the WASP team announced the discovery of 10 new hot Jupiters, found by transits and confirmed by Doppler spectroscopy. The 10 new planets, named WASP-6 b though WASP-15 b, had all been found in just six months of work. Ground-based transits were going wild, but what was CoRoT going to find?

April 3, 2008—Gordon Walker was at it again. He wrote a commentary in *Nature* about a new technique for improving the precision of Doppler planet searches, a technique that might enable ground-based astronomers to detect the Doppler wobble of planets as low in mass as Earth. Walker's commentary accompanied a paper by a group at the Harvard-Smithsonian Center for Astrophysics and MIT that had developed a new "laser comb" technique that could provide a reference standard for Doppler searches considerably better than the techniques currently in use. The new technique used a rapidly pulsed laser as the spectral reference source. The claim was that this laser comb would allow astronomers to measure the speeds of planet-host stars to a precision as small as 0.02 mile per hour (0.01 meter per second), a level about 100 times lower than the best current measurements.

Such a precision might make possible the detection of the Doppler wobble induced in a Sun-like star by an Earth-mass planet on an Earth-like orbit, but only if the host star happened to oblige. The surface of the Sun is covered with convective cells, like a pot of boiling water, with speeds on the order of miles per second (1.6 kilometers per second) that dwarf the periodic Doppler planet signals. Only if the host star's surface was sufficiently well behaved for all of these upwellings and downwellings to average out to something very close to zero could the laser comb technique make the revolutionary contribution it hoped to make. Star spots and stellar pulsations added yet other complications to the goal of reaching down to 0.02 mile per hour. Walker concluded that in theory, the laser comb could find Earths, but he cautioned that in the real world of noisy stars, "substantial hurdles" remained.

May 7, 2008—Michael Griffin announced that Ed Weiler had agreed to become the next long-term head of NASA's Science Mission Directorate. Weiler promised to keep the Mars Science Laboratory rover program alive and to cover its cost overruns by "spreading the pain." Would the pain be confined to the Mars exploration program, or would it affect NASA's entire planetary exploration program, or might it creep into NASA's other science programs as well? NASA's budget for planet finding beyond the Kepler Mission had already been diminished to close to the vanishing point, so for once it did not seem to be in peril, but the prospects for restoring the Navigator Program to fiscal health seemed remote in such a budget environment.

May 16–17, 2008—The Kepler Mission held a Science Team meeting at the Center for Astrophysics in Cambridge, Massachusetts. For once, the news was nothing but good. With just nine months remaining before launch, this was a tremendous relief. Bill Borucki announced that Kepler was essentially good to go. The entire telescope had been assembled, including the spacecraft, and subjected to a number of tests. These included "shake and bake," where the space telescope was first shaken to simulate the forces that it would experience during the Delta 2 launch from the Cape, and then placed in a large vacuum chamber, where it was heated up and cooled down to simulate the temperature variations it would encounter in interplanetary space. Kepler's instruments had performed flawlessly. In particular, Kepler's tests proved that it was able to measure the miniscule dimming in stellar brightness produced by a transiting Earth.

The only remaining concerns were the reaction wheels, which would be spun up or down in order to change the orientation of the telescope. Other NASA space telescopes with the same reaction wheels had developed problems with the wheels well before their design lifetimes had been exceeded. To maximize their lifetimes on orbit, Kepler's reaction wheels were removed from the spacecraft and sent back to the manufacturer to be rebuilt with better parts.

Kepler would be shipped to Cape Canaveral in December. Kepler had passed successfully into the phase of development called ATLO, which stands for Assembly, Test, and Launch Operations, a phase never before encountered by a NASA space telescope designed specifically to find extrasolar planets. This was a whole new world for NASA planet finding.

CoRoT was being suspiciously quiet. Two hot Jupiters had been announced since its launch in late 2006, but nothing else. The word was that the CoRoT team had decided to buy 40 nights of time on the Keck I telescope in order to use the HIRES spectrometer to confirm that some of the their most interesting transit detections really were exoplanets. Given the going rate of about $100,000 per night of Keck time, CoRoT was looking to spend $4 million to confirm their transit discoveries with Doppler wobbles. The Kepler project was planning on obtaining a similar number of Keck nights to confirm the discoveries that Kepler would be making in 2009.

May 19–23, 2008—Annie Baglin of the Paris Observatory, CoRoT's principal investigator, had the honor of announcing two more planet discoveries at an IAU Symposium on "Transiting Planets" in Cambridge,

Massachusetts. She showed that CoRoT had found two more hot Jupiters, named CoRoT-Exo-4 b and CoRoT-Exo-5 b, with masses slightly less than Jupiter and orbital periods of 9 and 4 days, respectively. A hot brown dwarf with a mass 20 times that of Jupiter, CoRoT-Exo-3 b, had also been found, with a peculiarly small radius. CoRoT now had four exoplanets in the bag, but nothing so far that did not appear to be a gas giant planet.

However, the CoRoT team hinted that they might have something remarkable, a possible transit signal that implied a planet with a radius about 1.7 times the radius of the Earth. If this materialized, CoRoT might have bagged its first hot super-Earth. Keck HIRES time would be needed to see exactly what was causing the dimming seen by CoRoT. Was it a super-Earth, or just a background eclipsing binary star masquerading as a planet?

May 29–30, 2008—JPL held "Exoplanet Forum 2008" at the Pasadena Hilton Hotel, a meeting intended to produce a series of recommendations about future planet-hunting missions that could be presented to the upcoming NAS Decadal Survey committees. Given NASA's budgetary straitjacket, the best that could be hoped for any time soon was a $600 million mission, perhaps SIM-Lite, perhaps something else. As a sign of the hard times, the name for JPL's planet-hunting program had been changed once again: the "Navigator Program" was now the "Exoplanet Exploration Program," or ExExP, and the Navigator Program's "NP" joined the dust pile of past NASA acronyms for the field. Perhaps ExExP would succeed where the others had failed.

The Kepler Mission would be moved from the Discovery Mission Program to ExExP once it was formally commissioned on orbit as a working space telescope in early 2009. At that point, ExExP would have its first working space telescope designed to find Earths. Kepler would effectively become NASA's first TPF telescope.

On Friday, the last day of the Exoplanet Forum, David Bennett told me that the microlensing teams had made another major discovery, but this time he did not reveal what it was. I would have to wait until Monday to find out.

June 2, 2008—The microlensers set a new record for the lowest-mass planet: the microlensing event MOA-2007-BLG-192 implied the existence of a cold super-Earth with a mass just 3.3 times that of Earth, orbiting at a distance of 0.62 AU from an object with a mass estimated at 6% of the Sun's mass—that is, a brown dwarf. Given the low mass of the central object, it was not clear whether this "planet" should be called a planet, but the fact that planetary-mass objects orbited even brown dwarfs was a strong indication that super-Earths could form just about anywhere there was a disk of raw material.

June 11, 2008—The IAU decided to tidy up some remaining business from the 2006 vote in Prague to demote Pluto from planethood to a new category of dwarf planets. Henceforth, Pluto and Eris would be officially known as plutoids, the IAU name for dwarf planets in the outer solar system. Ceres would remain as simply a dwarf planet, with no further laurels for its résumé.

June 12, 2008—The reaction to the IAU's plutoid decision was swift and harsh, as might be expected for such a high-stakes issue.

Mark Sykes and Alan Stern declared that the IAU's decision should be ignored, and Stern hinted that the IAU could be replaced with a new organization, in effect equating the IAU with the League of Nations, which was replaced by the United Nations after World War II. The Battle of Prague would be rejoined at a meeting to be held near Baltimore, Maryland, in August 2008. The Applied Physics Laboratory at Johns Hopkins would hold a "Great Planet Debate" and return to the Pluto fray, although this time, no votes were to be taken to decide the question of planethood.

June 16, 2008—The Swiss were at it again. At a meeting in Nantes, France, on super-Earths, Michel Mayor and his HARPS team announced that they had discovered a planetary system with not one, not two, but three super-Earths, with masses as low as 4.2, 6.9, and 9.2 Earth masses. These were more hot super-Earths, with orbital periods ranging from 4.3 days to 20.5 days, but the fact that there were three super-Earths in orbit around a single star made it clearer than ever that Mother Nature likes to make rocky planets. And what Mother Nature likes, Mother Nature gets, in spite of the fact that the host star for the triple-Earths system was metal-poor, with only half the rocky material of the Sun. The star, HD 40307, is about 40 light-years from Earth, making the three super-Earths prime candidates for future investigation by telescopes such as the TPFs and Darwin.

August 15, 2008—The Great Planet Debate was held at the Applied Physics Laboratory in Laurel, Maryland. The debate featured Neil Tyson and Mark Sykes continuing the public argument, begun in Prague 2 years earlier, over whether Pluto should be considered a planet. Sykes argued in favor of planethood for objects such as Ceres and Pluto, whereas Tyson argued against increasing the number of Solar System planets to 13 or more disparate bodies. No consensus was reached, setting the stage for a probable return to this question in 2009 at the IAU General Assembly in Rio de Janeiro.

August 20, 2008—Bill Borucki e-mailed the Kepler Science Team that the planned mid-February launch of Kepler would be delayed to no earlier than April 10, 2009, because of problems with the third stage of the Delta 2 rocket. Kepler was ready to launch, but until the third stage was judged safe, Kepler would have to sit in cold storage at the Cape, biding its time.

August 25, 2008—A member of the CoRoT science team sent out an e-mail, perhaps mistakenly, to the entire exoplanet community, addressed to the CoRoT team. The e-mail stated that the Kepler launch date had been postponed by at least several months and noted that "We have a little more breathing room. . . ." Evidently CoRoT could feel Kepler's hot breath on its neck in the race to find the first transiting Earth.

September 29, 2008—I read e-mail from Bill Borucki informing the Kepler Science Team that Kepler's launch date was now set for the night of March 4, 2009. The CoRoT science team lost a month of breathing room as a result. Kepler was good to go for launch.

On the same day, Ed Weiler announced that the SM4 mission to repair and upgrade Hubble would not take place as planned on October 14. Hubble had suffered another major equipment failure a few days earlier and was now unable to send any science data back to

Earth. SM4 would be delayed until February 2009 at the earliest, while Hubble's controllers tried to switch to the backup science data unit, and NASA pondered the problem of how best to fix Hubble. Kepler might beat SM4 to orbit.

October 2, 2008—The NASA Advisory Council was told that the Mars Science Laboratory was facing more cost overruns, with the total cost now well over $2 billion. This placed the JPL mission in the delicate position of facing a Cancellation Review, required for any mission with a cost overrun of 30% or more. Ed Weiler had another major headache on his hands. But rumors began to circulate that Weiler was gunning for Griffin's job, once the new president took office on January 20, 2009. Griffin might be handing the three envelopes to Weiler soon afterward.

January 4–8, 2009—The American Astronomical Society will hold its winter meeting in Long Beach, California, where the NAS 2010 Decadal Survey is expected to shift into high gear. The Survey chair and committee members will be presented with the astronomical community's desires for the next 10 years of big-ticket telescopes. The fate of SIM will be on the line. Although a high ranking by the Decadal Survey would be no more of a guarantee of success than that of the previous Surveys, a low ranking would probably sound the death knell for SIM.

February 2–5, 2009—The CoRoT Mission will hold its first major conference, in Paris. This meeting will highlight CoRoT's discoveries, both those already announced and those held back for the Paris meeting. What will CoRoT announce? Undoubtedly, CoRoT will ballyhoo a hot super-Earth or two, confirmed by ground-based Doppler spectroscopy, but what else will be announced? Perhaps a habitable super-Earth?

Spring 2009—The Kepler Mission will be launched into an Earth-trailing orbit from the Kennedy Space Center, at Cape Canaveral, Florida, on a Delta 2 launch vehicle. As a consequence of changes in the Air Force's reliance on Delta 2 vehicles to put spy satellites in orbit, the launch pad will be dismantled after Kepler's departure. Kepler will have the dubious honor of being the last spy satellite to be launched from that site, although in this case it is a satellite designed to spy on other planetary systems.

The year 2009 is the 400th anniversary of Kepler's publication of his first two laws of planetary motion and the 400th anniversary of Galileo Galilei's first use of the astronomical telescope. In addition, 2009 is the 200th anniversary of the birth of Charles Darwin and the 150th anniversary of the publication of his seminal book *On the Origin of Species by Means of Natural Selection*, the pivotal book for understanding the fossil record of life on Earth, the concept of evolution, and hence the origin of life on habitable worlds. Although Kepler's launch date had been originally planned for 2006, it was delayed 3 years, largely as a result of NASA headquarters actions. Bill Borucki joked that perhaps NASA headquarters had wanted to ensure that Kepler was launched in 2009 in order to honor these four major anniversaries.

One hour after launch, the wires from the Kepler space telescope to the third stage rocket will break off, and Kepler will begin its commissioning and on-orbit testing phase. The sensitive light detectors that will search for the tiny periodic dimming of stars caused by transiting Earths will be tested first with the telescope's dust cover still attached. In other words, Kepler will begin by seeing what it can see when there is nothing to see. Whatever it does see with the dust cover attached will need to be removed from what Kepler sees once its eye is open to the sky. After a first week of spacecraft and detector testing, the dust cover will be blown off, never to return; Kepler has no way to close its eye once it is opened. The Kepler Science Team will focus the telescope and start to calibrate the detectors against real stars, work that is expected to require several more weeks of testing and experimentation.

Launch plus one month, 2009—Provided that all has gone well with the previous month's work, Kepler will begin full-scale science operations, staring at a rich field of stars in the constellation Cygnus. The brightness of 170,000 target stars will be measured, averaged over 30-minute time periods, and downloaded by Kepler's radio transmitters to the Kepler Science Center at NASA Ames. The choice of the lucky 170,000 stars will be made from the list of 400,000 fairly bright stars in Kepler Input Catalogue, which are themselves a small subset of the 15,000,000 stars in the entire catalogue.

Within the first week, Kepler should start finding the periodic dimmings caused by hot Jupiters. Hot Jupiters occur around about 1% of solar-type stars, and about 10% of the hot Jupiters should be aligned with orbits that cause transits, so Kepler should have thousands of hot Jupiter candidates to sort through as soon as it gets started. Confirmation of the reality of any one of these hot Jupiter candidates will require Doppler observations, to make sure that the star is wobbling as expected. The Doppler confirmations will also provide the planet's mass, and with the planet's radius determined by the amount of the star's dimming during the transit, Kepler will revolutionize what is known about fluffy and dense hot Jupiters.

The Doppler follow-up work will be accomplished by the Keck HIRES spectrometer on Mauna Kea, and by a new spectrometer being built for use on the 168-inch (4.2-meter) William Herschel Telescope on La Palma, in the Canary Islands. The new spectrometer will be a clone of the Swiss HARPS spectrometer at La Silla, termed HARPS-NEF for "New Earth Facility." HARPS-NEF is being built by Harvard University, with the intention of achieving Doppler accuracy as good as 1 mile per hour (0.5 meter per second) with the standard Swiss thorium/argon reference cells, and maybe down to 0.2 mile per hour (0.1 meter per second) with the new-fangled laser comb technique. Such exquisitely small Doppler precisions are necessary to confirm the existence and mass of any hot and warm super-Earths found by Kepler.

After 3 years of transit observations, Kepler should begin to reveal the frequency of Earth-like planets on Earth-like orbits around stars like the Sun. Once the false positives are eliminated, and the Doppler follow-up work has been completed, NASA headquarters will hold a major press conference in its Webb Auditorium, and William Borucki will have the honor of telling the world just how frequently Earths occur. That is, of course, unless CoRoT makes the announcement first. Either way, after centuries—if not millennia—of speculation and wondering, we will finally know just how crowded the universe really is.

EPILOGUE

Why Don't You Ever Call?

Where are they?

—ENRICO FERMI (1950)

There is no democracy in physics. We can't say that some second-rate guy has as much right to opinion as Fermi.

—LUIS WALTER ALVAREZ (1967)

All the evidence gathered to date by over 10 years of planet hunting implies that Earth-like planets should be common in our neighborhood of the Milky Way Galaxy and, by inference, in other galaxies as well. But how common is common? Will essentially every nearby Sun-like star have a habitable world, or only 1 in 10, or 1 in 100, or 1 in 1000? The theoretical and observational discoveries of the last several decades support the prediction that the frequency of habitable worlds is more likely to fall in the upper end of this range than in the lower end. Given that Kepler will be able to produce a good measure of the frequency of Earth-like worlds, provided that this frequency is 5% or greater, there is every expectation that the Kepler Mission will succeed in determining this most basic parameter in any estimate of the prevalence of life in the universe. Even the CoRoT Mission has a good shot at discovering an Earth or two in such a crowded universe.

Physicist Enrico Fermi was equally positive about the likelihood of life beyond the Earth, yet he asked a related question that became known as Fermi's Paradox: If intelligent life exists elsewhere in the Galaxy, why haven't we heard from it yet? During the summer of the year that Fermi asked his famous question, George Wetherill studied physics under Fermi at the University of Chicago. Fermi would march into a hot classroom in suit and tie, and by the end of his vigorous 3 hours of lecturing, he would have taken off his tie, then his coat, and finally his dress shirt, leaving on only his undershirt and pants. The sight of the 1938 Nobel Prize winner in a damp undershirt made a lasting impression on the young Wetherill, who religiously maintained Fermi's high sartorial standards but avoided similarly sweaty situations. Perhaps Fermi's Paradox was discussed in class that summer, perhaps not, but Fermi and Wetherill were physicists who evolved into two of the world's first astrobiologists—scientists who seek to understand the genesis, evolution, and prevalence of life in the universe.

This book has not addressed the question of how microscopic life originates on habitable planets, much less the intelligent life of Fermi's Paradox, but has instead taken the position that "if you build it, they will come." That is, if habitable worlds are common, what is to prevent their hosting the evolution of some sort of primitive life forms over their billions of years of existence? Not every planet needs to witness the evolution of *Homo sapiens* for life to be considered a universal trait; methanogenic bacteria would do quite nicely, provided that they are capable of generating enough methane to serve as a biomarker for some future space telescope. Still, if microscopic and even macroscopic life is commonplace, then considering the billions of habitable worlds in the Milky Way, one must conclude that Earth is not likely to be the only planet where intelligent life arises.

So where are they? There are many possible answers to Fermi's Paradox, ranging from the practical, through the merely depressing, to the near-suicidal.

The fact that we have no proof of intelligent life elsewhere in the universe may simply mean that intelligent civilizations have all too finite lifetimes—that intelligent civilizations are born to die, just like individual human beings. Part of the burden of consciousness and of being alive is the knowledge that someday one will die. There is no reason to believe that the same sobering prospect should not apply to civilizations in general, just as it did to civilizations and evil empires in the past, from the Babylonians to the Third Reich. The list of possible threats to human civilization is long and seemingly growing. It includes (but is not limited to) global warming, toxic pollution, depletion of natural resources, mutual annihilation, lethal viruses lurking unseen in the Amazon River basin, and such external threats as incoming asteroids and comets, a nearby supernova, and a powerful gamma ray burst that happens to be directed right at the Solar System.

The next closest habitable world may lie only a few light-years away, but the closest intelligent civilization may be a thousand times more distant if intelligent civilizations are somewhat rare in the galaxy. If intelligent civilizations themselves last only thousands of years, there might be no time to drop by for a visit before the lights go out, even if travel at the speed of light were possible. It is not. Accelerating a rocket from rest to the speed of light would require an infinite amount of energy. Accelerating a rocket to even a small fraction of the speed of light requires far more energy than is conceivably available for rocket propulsion in anything other than the world of science fiction. The New Horizons Mission to Pluto will cover a distance of 30 AU in about 10 years; it would require another 100,000 years to travel to the closest star system, Alpha Centauri. There is little chance that an alien spacecraft is going to drop in on Earth any time soon.

Some argue that we should not send out strong radio pulses defiantly announcing our existence for fear that something else will hear us and come here to kill or enslave us—or maybe just to pick up some Starbucks and have a talk after a long ride. Perhaps the lack of radio signals from other worlds means the Galaxy is populated with frightened beings anxiously awaiting the scream of their interstellar air-raid sirens. The physics of interstellar travel eliminates the need for such worries.

Perhaps intelligent life survives but evolves into some new species that has no interest in colonizing the Galaxy or in directing interstellar beacons of radio waves in our direction. With the knowledge gained over the next several thousand years of scientific research, the natural curiosity of human beings might well be sated. Perhaps civilizations die of sheer boredom. What would it mean for a human being to live for a thousand years rather than for less than a hundred? How many bagels do you really want to eat? What would it mean for a civilization to last 10,000 years? A million years? A billion years? These are the natural time scales of the universe. Intelligent civilizations may just grow old and tired and lapse into a somnambulist state. Where are they? Everywhere, but we cannot hear them snore.

Even if life surely exists elsewhere in the Milky Way Galaxy and throughout the universe, the closest habitable worlds are likely to be in a phase of the development of life that is either pre-intelligence or post-intelligence. But this likelihood by no means detracts from our strong desire to search for such life. The former will tell us about our past, and maybe about how we originated and evolved, whereas the latter may tell us about our future. Finding a world of methanogenic bacteria would be just as mesmerizing as finding an Earth-like planet that has been repopulated by a thermophilic species better adapted to the runaway greenhouse heating caused by the short-sighted inhabitants who went before.

LIST OF ACRONYMS AND ABBREVIATIONS

ACS	Advanced Camera for Surveys
ATLO	Assembly, Test, and Launch Operations
AU	astronomical unit (average Earth–Sun distance of 93,000,000 miles)
CFHT	Canadian-France-Hawaii Telescope
CNES	Centre National d'Etudes Spatiales
CoRoT	Convection, Rotation, and Planetary Transits
ExExP	Exoplanet Exploration Program
ESA	European Space Agency
ExNPS	Exploration of Neighboring Planetary Systems
FAA	Federal Aviation Administration
FRESIP	Frequency of Earth-Sized Inner Planets
GAIA	Global Astrometric Interferometer for Astrophysics
HARPS	High Accuracy Radial Velocity Planet Searcher
HAT	Hungarian-made Automated Telescope
Hipparcos	High Precision Parallax Collecting Satellite
HIRES	High Resolution Echelle Spectrometer, Keck Observatory, Hawaii
IAU	International Astronomical Union
ICBM	intercontinental ballistic missile
JPL	Jet Propulsion Laboratory, Pasadena, California
MER	Mars Exploration Rover
MicroFUN	Microlensing Follow-Up Network
MOA	Microlensing Observations in Astrophysics
MOST	Microvariability and Oscillations of Stars Telescope
MPF	Microlensing Planet Finder
NAS	National Academy of Sciences
NASA	National Aeronautics and Space Administration

NEF	New Earth Facility
NICMOS	Near Infrared Camera and Multi-Object Spectrometer, Hubble Space Telescope
NORAD	U.S. North American Aerospace Defense Command
NP	Navigator Program
OGLE	Optical Gravitational Lensing Experiment
OSI	Orbiting Stellar Interferometer
PLANET	Probing Lensing Anomalies Network
SIM	Space Interferometry Mission
SIM-PH	Space Interferometry Mission–Planet Hunter
SM	Servicing Mission
SMD	Science Mission Directorate (NASA)
SOFIA	Stratospheric Observatory for Infrared Astronomy
SST	Spitzer Space Telescope
TOPS	Towards Other Planetary Systems
TPF	Terrestrial Planet Finder
TrES	Trans-Atlantic Exoplanet Survey
UBC	University of British Columbia
UC	University of California
UCAR	University Corporation for Atmospheric Research
USNO	U.S. Naval Observatory
WASP	Wide Angle Search for Planets

INDEX